Mechanical Properties of Xenes

David J. Fisher

Published by **Materials Research Forum LLC**
Millersville, PA 17551, USA

Published as part of the book series
Materials Research Foundations
Volume 176 (2025)
ISSN 2471-8890 (Print)
ISSN 2471-8904 (Online)

Print ISBN 978-1-64490-364-3
ePDF ISBN 978-1-64490-365-0

Distributed worldwide by

Materials Research Forum LLC
105 Springdale Lane
Millersville, PA 17551
USA
http://www.mrforum.com

Printed in the United States of America
10 9 8 7 6 5 4 3 2 1

Table of Contents

Introduction 1

Graphene 2

    Moduli 2

    shear 3

    Young's 3

    Tensile strength 6

    Toughness 10

    Fatigue 11

Silicene 12

    Mechanical Properties 14

    Stone-Wales Defect 42

Germanene 45

    Preparation 45

    Structure 48

    Mechanical Properties 52

    Surface Interaction 53

    Defects 60

Stanene (Tinene) 61

    Structure 62

    Mechanical Properties 63

    Defects 67

Antimonene 68

    Preparation 68

    Mechanical Properties 71

Indiene 74

    Mechanical Properties 74

Arsenene 75

    Mechanical Properties 76

Phosphorene 85

    Preparation 85

Structure 87
Mechanical Properties 88
Defects 97
Bismuthene **100**
Mechanical Properties 100
Borophene **106**
Structure 106
Mechanical Properties 108
Plumbene **115**
About the Author **117**
References **118**

Introduction

The discovery of graphene began a new era in materials science. Graphene is a 2-dimensional planar honeycomb array of carbon atoms in sp^2-hybridized states. It was natural to wonder whether other elements could form graphene-like structures. The answer was affirmative, and there now exists a menagerie of analogous materials. Many of them are of interest largely because of their electronic properties but, like graphene which is famed for its great strength, it is useful to catalogue here the mechanical properties of the analogues although, admittedly, much of the information is based upon computer modelling rather than experimental investigation. Density-functional theory methods are a particularly powerful tool when investigating the electronic structure of nanomaterials. By using this method, the many graphene-like materials have been explored. Monolayer materials in general have been widely studied and, those which have been created include group-IV elements (silicene, germanene, stanene) and group-V elements (phosphorene). The broad range of band-gaps offered by 2-dimensional monolayers also makes arsenene, bismuthene and antimonene of great interest. Their mechanical properties are surveyed in the present book, with the contents being limited to the pure element, apart from occasionally considering the involvement of hydrogen which turns the 'ene into an ane'.

Materials Research Forum LLC

https://doi.org/10.21741/9781644903650

Graphene

Moduli

Tensioned graphene membranes are generally assumed to have a stretching modulus of some 340N/m, and to possess a negative temperature-independent thermal expansion coefficient which is due to transverse phonon modes. The 2-dimensional Young's modulus and thermal expansion were determined[1] as functions of temperature by using laser interferometry to detect displacements of the membrane in a cryostat. The modulus decreased markedly with increasing temperature, leading to a positive temperature-dependent thermal expansion coefficient. Surface contaminants were suggested to explain this behaviour. Nano-indentation experiments and simulations have been used[2] to estimate the elastic modulus of freely-suspended multilayer graphene. Interlayer sliding could affect the estimation of the modulus. Molecular dynamics simulations demonstrated that the estimated modulus depended upon the level of interlayer sliding which was involved. Under fully-clamped boundary conditions, which prevented interlayer sliding, the modulus was constant, regardless of the number of layers. Under weakly-clamped boundary conditions which allowed interlayer sliding, the modulus decreased with increasing number of layers. Again in the case of weakly-clamped conditions, a few wrinkles formed within the interlayer and coalesced into a single large wrinkle via interlayer sliding.

Atomic force microscopy based property measurements revealed[3] a clear defect density dependent behaviour of the elastic modulus of graphene on a substrate when the defect density was greater than $10^{12}/cm^2$, whereupon a clear decay was observed in the stiffness of graphene. The defect-density dependence was attributed to the appearance of amorphous graphene.

The effects of Stone-Wales defects upon the elastic modulus and vibrational responses of single-layer graphene sheets were determined[4]. The atomic positions of carbon atoms within the defective layers were used as initial conditions for finite-element calculations of the mechanical properties and natural frequencies of carbon monolayers. Two sets of initial atomic positions were used: the harmonic solution predicted by discrete dislocation theory and a fully-relaxed configuration. The mechanical properties were based upon the equivalence between the interatomic potential energy of the atomic model and the strain energy of the continuum model. An extension of the classical model accurately defined the mechanical properties in terms of the length of each atomic bond.

shear

Interlayer shear deformation occurs during the bending of multi-layer graphene with unconstrained ends, and affects its bending rigidity. An investigation was made[5] of the bending stiffness and interlayer shear modulus of few-layer graphene by considering its self-folding conformation, on a flat substrate, by using atomic force microscopy and non-linear mechanics modelling. The bending stiffness of 2- to 6-layer graphene obeyed square-power dependence upon thickness and the interlayer shear modulus range from 0.36 to 0.49GPa. Weak interlayer-shear interaction was concluded to exert an appreciable stiffening effect upon multi-layer graphene. The evolutionary approach to multi-gene genetic programming was used[6] to study the effects of aspect ratio, temperature, number of atomic planes and vacancy defects upon the tensile and shear moduli of monolayer graphene sheet. The model was able to predict the engineering moduli of armchair and zigzag-oriented graphene sheets. The number of defects present had the greatest effect upon the moduli.

The out-of-plane shear modulus of few-layer graphene was measured[7] by using a blister test. From the deflection profile of the blister, it was deduced that the average shear modulus was 0.97GPa. Internal friction and shear-modulus measurements were made[8] of various types of synthesized graphene film. The films were transferred to a water bath, and then re-deposited on a high-Q monocrystalline silicon mechanical oscillator. A minimum thickness dependence of internal friction and shear modulus was found, for chemical vapour deposited graphene films on nickel, which ranged from 6 to 8nm. The shear modulus of the multi-layered films averaged 53GPa. The internal friction exhibited a temperature-independent plateau below 10K. The plateau was as high as the so-called glassy range within which tunnelling states dominate the internal friction of amorphous solids. Chemical vapour deposited graphene films on copper foil were 90 to 95% single-layer. The shear moduli of such single-layer films were some 5 five times higher, with an average value of 280GPa, and the low-temperature internal friction was undetectable.

Young's

It has been pointed out that the equivalent Young's modulus and thickness of graphene sheets should correspond to stretching and bending conditions. These features were explored[9] by using an analytical method which was based upon atomic interactions in a continuum elastic model. The results unified the modulus for the cases of stretching and bending, and also determined the matching thickness. The equivalent modulus and the thickness of the sheet were deduced to be 2.81TPa and 1.27Å, respectively. The use of classical molecular dynamics simulations established a relationship between thermally excited ripples and the Young's modulus of defective graphene sheet[10]. The presence of out-of-plane intrinsic ripples stabilized the graphene membranes. The presence of vacancy and Stone-Wales defects reduced the Young's modulus of the sheets. Graphene sheet with vacancy defects had a higher Young's modulus than that of a sheet with Stone-Wales defects. The room-temperature Young's modulus of pristine and defective graphene sheet was about 1TPa. The effect of vacancy and Stone-Wales defects upon the Young's modulus of graphene sheets was investigated[11] by using molecular dynamics simulations. These showed that defects reduced the Young's modulus of graphene sheets, and that the modulus decreased with increasing number of defects. Vacancy defects decreased the modulus, but their reconstruction was an important factor in stabilizing the modulus. The hydrogenation of vacancy defects could increase the Young's modulus of defective graphene, but the hydrogenation of Stone-Wales defects had the opposite effect.

The size-dependent Young's modulus and carbon-carbon binding energy in an homologous series of graphene nano-ribbons were determined[12] by using electron first-principles computations. A linearity between the binding energy and modulus was observed, thus making it possible to predict the size-dependent Young's modulus of the nano-ribbons via a single-point energy calculation of the nano-ribbon ground state. A quantitative relationship was derived which related the Young's modulus to the total energy and the number of carbon atoms within the ribbon. In the limit of extended graphene sheets, the Young's modulus was 1.09TPa.

The Young's modulus of graphene with various rectangular and circular vacancy defects was determined[13] by means of molecular dynamics simulations. The modulus was mainly related to the size of vacancy defects perpendicular to the stretching direction. A linear reduction in the modulus with an increasing concentration of mono-atomic vacancy defects occurred. A blunting effect of vacancy edges was demonstrated, and the crack-tip radius was 4.44Å.

The Young's modulus of graphene was estimated[14] by measuring the strain which was applied by a pressure difference acting across graphene membranes. The strain which was induced in pressurized graphene balloons could be estimated directly from the peak shift of the Raman G-band. By comparing the measured strain with numerical simulations, the moduli of single- and bi-layer graphene were found to be 2.4 and 2.0TPa, respectively.

Molecular dynamics methods were used[15] to study the Young's modulus of monolayer to 19-layer graphene by using nano-indentation and tensile theory. The number of layers had a marked effect upon the modulus when using nano-indentation theory, but the modulus deduced from tensile theory was insensitive to the number of layers. The multilayer graphene transformed into bulk graphite with increasing number of layers, indicating that nano-indentation theory had reached its limit. Simulation of nano-indentation for circular membranes of various radii determined the exact threshold layer number for applying nano-indentation theory. The nano-indentation of monolayer graphene was studied[16] by means of molecular dynamics simulations. The response of the graphene to indentation was deflection dependent. At small deflections, the response obeyed a point-load model. Large-deflection indentation obeyed the sphere-load model. By using differing response models in different deflection ranges, a consistent Young's modulus was obtained. The calculated Young's modulus was about 1.00TPa.

A new method was proposed[17] for constructing a reduced-order model for the dynamics of thermalized graphene membranes. An orthogonal decomposition algorithm based upon time responses deduced from molecular dynamics simulations was used to construct a set of orthogonal modes which were employed to build a reduced-order model. This model was used to study the non-linear dynamics of graphene membranes at large amplitudes in order to deduce the Young's modulus. The latter modulus incorporated the effects of nano-scale thermally induced dynamic ripples and was temperature- and size-dependent.

Vacancy defects in graphene have an effect upon the intrinsic in-plane strength. The equivalent Young's modulus was deduced[18] from the strain energy as being an important factor in evaluating the stiffness of graphene on the basis of molecular theory. The results showed that centre-concentrated vacancy defects impaired the elastic stiffness under shear stresses. The effects of periodic and regular vacancy defects depended upon the boundary conditions. When compared with the results for graphene with vacancy defects under uniaxial tension, the damaging effects of vacancy defects were less evident in graphene under shear stresses. Tight-binding atomistic simulations were used[19] to study native defects in graphene. Their formation energy could be as high as several electron volts. Defects could introduce large deformation and stress fields which had a spatial distribution that was closely related to their symmetry. They were suggested to define the basin of mutual interaction between two neighbouring defects. Defects affected differently the linear elastic moduli of monolayer graphene. The Young's modulus and Poisson ratio decreased, but their dependence upon the defect surface-density was markedly greater for vacancy-like than for number-like defects.

Carbon nanotubes can be regarded as being rolled-up graphene, so any formulation which describes the behaviour of the nanotubes must incorporate information related to both the initial graphene configuration and the final nanotube configuration. A simplifying assumption is that the length and perimeter of the nanotube equal the corresponding dimensions of the unrolled initial configuration, neglecting induced hoop and longitudinal strains. A new and purely non-linear continuum model[20] was suitable for describing the large deformation of the graphene without requiring the simplifying assumption. Closed-form expressions for the Young's modulus and critical buckling strain of single-walled carbon nanotubes were found to be functions of the elastic constants of the graphene, geometrical properties of the tube section and a parameter which depended upon the chirality angle. This parameter was the coefficient of the non-linear term.

Molecular dynamics simulations of multilayer graphene were used[21] to investigate the effect of deflection, layer-number, temperature and stack pattern upon the determination of the Young's modulus by using nano-indentation models. The modulus was inconsistent under differing strain conditions, with its value being much higher for small deflections. The modulus values of multilayer graphene were quite consistent, regardless of the stack pattern. The relationship between modulus and temperature was almost linear at 300 to 1000K, with the modulus increasing in proportion to the temperature. The Young's modulus was about 1.00TPa for multilayer graphene. This predicted value was in good agreement with that for monolayer graphene and was close to measured values.

The Young's modulus of graphene was investigated[22] via the intrinsic thermal vibrations in graphene which are monitored by molecular dynamics. This was used to show that the modulus increased with increasing size, and saturated above a threshold size. It increased from 0.95 to 1.1TPa as the temperature was varied between 100 and 500K. It was insensitive to the isotopic disorder in the low-disorder region (<5%) and decreased gradually upon further increasing the percentage of disorder.

Tensile strength

The tensile strength, fracture strain and elastic modulus of graphene sheet which contained randomly distributed vacancies were investigated[23] by means of molecular dynamics simulations. The presence of defects markedly impaired the mechanical properties of the sheet. When the number of defects was sufficiently high, the failure mechanism changed. The defects impaired the mechanical properties by reducing the tensile strength, failure strain and elastic modulus, but the differences between 0.01K and 300K, 600K or 900K remained almost unchanged as the percentage of vacancies increased. A detailed study was made[24] of the tensile strength and fracture strain, of monolayer graphene which contained common topological defects and nanopores, by using molecular dynamics and atomic-scale finite-element methods. The tensile strength exhibited a complex dependence upon the defect-shape, size and chirality. Some defect geometries were mechanically superior to other geometries, thus encouraging the topological design of graphene so as to optimize its properties.

By conducting atomic-scale characterization and *in situ* high-field tests in a field-ion microscope, it was shown[25] that the inherent strength of graphene is 99.34GPa and represents the intrinsic tensile strength of statistically defect-free graphene nanoribbons. Mechanical failure of graphene ribbons under tensile stress when near to the ideal strength determined the fundamental upper limit on the internal stress in electromechanical graphene devices.

Molecular dynamics and an adaptive intermolecular reactive bond order potential was used[26] to determine the effect of 5-8-5 defects (divacancies associated with pentagon-octagon-pentagon configurations of carbon atoms) upon the tensile strength of graphene sheets. Computer simulations revealed a multi-stage behaviour of the fracture process in sheets which contained 5-8-5 defects. The final stage of fracture involved the formation of carbon monatomic chains that joined separate parts of graphene sheets under mechanical loading. The tensile strength was markedly decreased by tens of percent due to the 5-8-5 defects.

Molecular dynamics simulations with an adaptive intermolecular reactive empirical bond order potential were again used[27] to investigate the effects of rectangular nanoholes having various areas, aspect ratios and orientations upon the tensile strength of defective graphene. The simulations showed that varying the area, aspect-ratio and orientation of rectangular nanoholes could greatly affect the tensile strength of defective graphene. Material with a larger area of rectangular nanoholes exhibited a larger fall in tensile strength. The tensile strength of armchair and zigzag-edged graphene decreased monotonically with increases in the area of rectangular nanoholes. Changes in the aspect-ratio and orientation of rectangular nanohole could decrease or increase the tensile strength of defective graphene; depending upon the tensile direction. The tensile strength of defective graphene with a large area of nanohole was more sensitive to changes in aspect-ratio and orientation than was defective graphene having a small area of nanohole. The variation in tensile strength of defective graphene which was predicted by molecular dynamics simulations was in good agreement with predictions arising from energy-based quantized fracture mechanics. The results suggested that the effect of nanoholes upon the tensile strength of graphene furnished essential information for the optimization of mechanical properties.

Molecular dynamics simulations with the adaptive intermolecular reactive empirical bond order potential were used[28] to study the effect of temperature and orientation upon the tensile strength of pristine and defective bilayer graphene sheet. The decrease in the tensile strength of pristine AA stacked material, due to the presence of a vacancy, was appreciable at 300K but decreased at 1073K. This behaviour was reversed in the case of

AB stacked material, with the percentage loss of strength at higher temperatures due to defects being greater than that at room temperature. It was suggested that this phenomenon occurred when defects were present in the armchair direction. In the case of AB stacked material, the zigzag direction was more defect-tolerant at room and high temperatures. The variation in tensile strength due to orientation was in good agreement with predictions arising from potential energy concepts.

Molecular dynamics simulations showed[29] that the strength which is measured in uniaxial simulations and the strength which is estimated from nano-indentation fracture can be very different. Fracture in tensile loading occurred at the same time as the onset of crack nucleation near to 5-7 defects. Graphene sheets meanwhile often support indentation loads following crack initiation because a sharply concentrated stress near to the tip does not offer a sufficient driving force for further crack propagation. Due to the concentrated stress, the estimation of strength is sensitive to the indenter-tip position along the grain boundaries. It approaches the strength of pristine graphene if the tip is located slightly apart from the grain boundary line. This revealed the limits of nano-indentation testing for quantifying the strength of graphene, and showed that a loading method which matched the failure mechanism had to be chosen.

Higher temperatures lead to higher atomic kinetic energies and to easier failure of graphene, and just a one-atom vacancy can lead to a sharp reduction in strength. Analytical expressions associated the tensile strength of monolayer graphene with temperature and with the length of a centrally-positioned straight crack. Molecular dynamics simulations were used[30] to compute necessary numerical data which were then fitted to non-linear regression surfaces. The non-linear analytical equations were able to predict the strength of graphene, given the chirality, the temperature and the size of the centre crack.

Large-scale ultra-light nanoporous graphene having a 3-dimensional bi-continuous nano-architecture offered an orders-of-magnitude higher strength and elastic modulus than did then-known ultra-light carbon materials under compression or tension. The high-strength nanoporous graphene also exhibited[31] a high tensile ductility and work-hardening. The excellent mechanical properties of the nanoporous graphene arose from seamless graphene sheets in the bi-continuous nanoporosity which preserved the intrinsic strength of atomically thick graphene in the 3-dimensional cellular nano-architecture. Density functional theory calculations indicated[32] that, during the early stages of uniaxial tensioning of graphene allotropes with high Poisson ratios, bond-rotation effectively reduced bond-stretching and increased the fracture strain. During the deformation of a Gr10 allotrope with a Poisson ratio of 0.8, bond-rotation led to an S-shaped stress-strain curve like that of an elastomer. The tensile strength and Young's modulus of the Gr10 exceeded the theoretical cohesive strength of E/10, and attained E/1.7. A universal relationship between bond-lengths and charge-density distribution along bond paths was found to apply to all carbon-carbon covalent bonds, and all carbon-carbon bonds thus obeyed a common form of bond-force versus bond-strain curve, such that the strength, S, and the bond-stiffness, K, were related by S ≈ K/9. The E/10 result thus remained valid for low Poisson-ratio graphene allotropes when deformation is dominated by bond-stretching. It was suggest the compromise between bond-stretching and bond-rotation could be used to enhance the fracture strain of 2-dimensional carbon structures.

A combined on-chip tension and cracking method were used[33] to determine the fracture toughness and strength of monolayer monodomain free-standing graphene. Crack-arrest tests revealed a fracture toughness of 4.4MPam. Tension on-chip indicated a Young's modulus of 950GPa, a fracture strain of 11% and a tensile strength of up to 110GPa. A record-breaking stored elastic energy some $6GJ/m^3$ was confirmed by thermodynamics and by quantized fracture mechanics. A circa 1.4nm crack size was found to be responsible for graphene failure, and was linked to 5-7 pair defects. The effect of the a/W ratio upon the fracture toughness of graphene was investigated[34] via molecular dynamics simulations by using an hexagonal lattice structure for the monolayer graphene sheet. Two different square domains with 10 different lattice orientations and 13 different initial crack lengths were considered. The lower bound on the a/W ratio was defined to be the value beyond which the Young's modulus began to decrease with increasing a/W. The fracture toughness increased with increasing a/W ratio until it reached a peak value before decreasing with further increases in the a/W ratio. The upper bound on a/W was recommended to be the value beyond which fracture toughness started to decrease. The a/W range of 0.15 to 0.45 was recommended for the prediction of the fracture toughness of graphene.

The fracture properties of stacked bilayer graphene sheets were studied[35]. The known elastic modulus and the known displacement boundary conditions at the onset of crack growth were used to estimate the far-field force and stress. The material was markedly tougher than existing mono- or multi-layer graphene, with a mode-I fracture toughness of about 28.06MPa.

Penta-graphene is a carbon allotrope having a 2-dimensional arrangement of carbon atoms in a pentagonal tiling. The monolayer is mechanically and thermodynamically stable, with a band-gap of 3.25eV. The effect of temperature on the mechanical properties of penta-graphene monolayers at up to 2000K was determined[36], and the results showed that the strain-rate had a marked effect upon the mechanical properties, with a reduction of 67% in the ultimate tensile strength of 35.88 to 11.83GPa.nm and a Young's modulus of 227.15 to 154.76GPa.nm.

Toughness

A continuum model which was originally proposed for estimating the elastic stiffness of nanomaterials was modified[37] in order to estimate the ultimate tensile strength and mode-I/mode-II fracture toughness of graphene. By assuming a modified Morse potential function and applying a uniform strain field for a perfect specimen, a non-linear stress-strain diagram could be plotted in order to estimate the tensile strength. In order to estimate fracture toughness, a strain-intensity factor was introduced and the near-tip solution of linear elastic fracture mechanics was written in terms of strain intensity factor to locate atoms of the cracked specimen. With the changes in bond distance and bond angle between atoms being set in the deformed state, the potential energy within a representative volume was treated as being the strain energy within the cracked specimen. The results which were estimated by using this model fell within a reasonable range.

A method was presented for the measurement of the fracture properties of graphene with various numbers of layers[38]. The method involved a custom tensile-tester and a repeatable process which permitted the transfer of patterned graphene structures. The fracture properties of graphene samples having various numbers of layers were tested, and the critical strain energy release-rate of fracture was 22, 71.3 and 145J/m^2 for 7, 12 and 15 layers, respectively. The increase in fracture toughness with that number was attributed to the need of brittle cracks to propagate through multiple weakly-connected layers. The steps involved during the fracture of graphene were studied[39] by performing *in situ* tensile tests. Small cracks which nucleated from the graphene edges were deflected into major cracks with localised kinks due to changes in the stress-intensity factor during crack advance. A finite-element analysis and the maximum energy release rate criterion were used to model the process. There was a slight degree of anisotropy in the fracture toughness, such that the armchair/zigzag ratio was 0.94.

Crack propagation along grain boundaries in polycrystalline graphene, as an ordered array of non-hexagonal defects, was analyzed[40]. The boundaries were considered to be the most favourable paths for crack propagation, because the carbon atoms along the boundaries experienced less-ordered interatomic interactions than did those in pristine graphene. When mixed-mode loading was applied to a crack which ran along a graphene boundary it was noted that, as the mode-mix increased, the fracture toughness of grain boundaries in graphene gradually increased. If the mode-mix was greater than 8°, the fracture toughness gradually decreased. This was thought to be unique to grain boundaries in graphene. The abnormally low fracture toughness of the boundaries in graphene for a high mode-mix was attributed to a competition between bond-rotation and bond-breaking mechanisms for carbon atoms, in conjunction with Stone-Wales transformation and nonagon structures at the crack tip along the grain boundaries.

A method was proposed[41] for the study of the fracture characteristics of graphene by using Raman spectroscopy and chemical-vapour deposited monolayer graphene with inserted cracks. The dynamic fracture process was experimentally monitored, and the fracture toughness was deduced by using Griffith's criterion and a strain distribution which was estimated from frequency shifts of the Raman bands. The fracture toughness parameters of 6.1MPam and $37.4J/m^2$ were comparable with previously reported theoretical and experimental values.

The interface adhesion toughness between multilayer graphene films was considered. The results of a circular blister test exhibited[42] an anomalous behaviour in which the adhesion toughness depended upon the number of graphene layers. It was shown that interlayer shearing and sliding near to the blister crack tip, caused by a transition from membrane-stretching to combined bending, stretching and through-thickness shearing, decreased the fracture-mode mix and led to a lower adhesion toughness. For a silicon oxide substrate and pressure-loading, the mode-mix decreased from 232% for monolayer films to 130% for multi-layer films. This caused the adhesion toughness to decrease from 0.424 to $0.365J/m^2$. The mode-I and mode-II adhesion toughness values were 0.230 and $0.666J/m^2$, respectively. With point-loading, the mode mix decreased from 741% for monolayer films to 262% for multi-layer films, while the adhesion toughness decreased from 0.543 to $0.438J/m^2$.

Due to their interconnected structure, and a synergistic effect of graphene ribbons and sheets, graphene hybrid fibres had[43] a strength of 223MPa. The graphene hybrid fibre also exhibited a toughness of $30MJ/m^3$. Dry-spinning was used[44] to fabricate continuous graphene fibres which exhibited a good toughness and flexibility. The dry-spun fibres had a toughness of up to $19.12MJ/m^3$.

Fatigue

A fatigue study was made[45] of free-standing 2-dimensional materials such as graphene. By using atomic force microscopy, monolayer and few-layer graphene was shown to exhibit a fatigue life of more than 10^9 cycles at a mean stress of 71GPa. The fatigue failure in monolayer graphene was global and catastrophic, with no progressive damage. Molecular dynamics simulations revealed that this was preceded by stress-mediated bond reconfiguration near to defect sites. Polycrystalline graphene nanoribbons tend to suffer brittle fracture before failure, and the irreversible breaking occurs at grain boundaries.

Materials Research Foundations **176** (2025)   https://doi.org/10.21741/9781644903650

Molecular dynamics simulations were used[46] to study the fatigue properties of bi-crystal graphene nanoribbons under periodic in-plane compression. Ribbon edges, rather than grain boundaries, dominated dynamic stability when the ribbon-width was smaller than 5nm. The fatigue failure was closely related to the misorientation angle between two grains because of the edge energy. It always began at the edges with higher energy and resulted in localized damage at the edges.

Graphene/polymer interfaces are constantly subjected to dynamic loading during use. *In situ* cyclic loading was used[47] to observe interfacial fatigue propagation at a graphene/polymer interface. It was shown to obey a modified Paris law. Cyclic loading via van der Waals contact was able to provoke fatigue fracture, even in pristine graphene, via a combined in-plane shear and out-of-plane tearing mechanism. Shear fracture was initiated mainly at fold junctions which were induced by cyclic loading, and then propagated parallel to the loading direction. A fracture-mechanics analysis was used to explain the kinetics of an exotic self-tearing behaviour of graphene during cyclic loading.

Silicene

Silicene was theoretically predicted in 1994 and was created experimentally much later, in spite of some skepticism[48]. Like graphene, silicene (figure 1) possesses properties that can be exploited in nano-electronic applications, such as Dirac electronic dispersion. There is also an ability to open a band-gap in the presence of an electric field or when deposited onto a substrate. Silicene is nevertheless very different to graphene with regard to its stability, atomic structure and processing but, due to its superior properties, it is a potential reinforcing agent. There once existed some doubt over whether graphene could be prepared, given that it contradicted the Landau-Peierls-Mermin-Wagner predictions that there could exist no stable flat form of such a structure[49]. There were similar doubts concerning silicene. As far back as 1994, the possibility of corrugation of the silicon analogue of graphite had already been theoretically considered. There was very little study of silicene before 2009, when its low-buckle structure was proved to be dynamically stable by using *ab initio* calculations. The buckling amounts to 0.454Å. Silicene is stable due to the transverse short-range displacements of certain atoms. The distortions are small and form various patterns. With decreasing temperature, transitions from disorder to order and from order to disorder can occur.

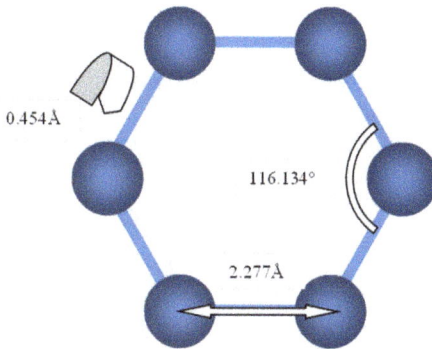

Figure 1. Basic silicene hexagon, showing Si-Si bond-length, bond-angle and buckling

Molecular dynamics simulation, using the Lennard-Jones potential, of a 2-dimensional array of silicon atoms at various temperatures and densities shows that the radial distribution function resembles the corresponding (111) surface of the face-centred cubic structure. For silicon, which is usually sp^3-hybridized, silicene is an unusual structure. Its structure was calculated as a possible candidate for nanoribbons of silicon grown onto the anisotropic Ag(110) surface. These nano-ribbons consisted of an arched honeycomb structure. Unlike silicon surfaces, which reacted strongly with oxygen, the silicene nano-ribbons were resistant to attack. Unlike graphene, silicene sheets are stable only if the small degree of buckling is present. Silicene is symmetrically buckled in each of its 6-membered units, and the buckling is periodically repeated across the surface. The symmetrical buckling of the 6-membered rings distinguishes it from graphene. A pseudo-Jahn-Teller distortion breaks the symmetry and leads to the buckling. Hydrogenation is much easier in silicene and the resultant silicane formation opens the band-gap and increases the puckering angle. Silicane will not be extensively considered here because it does not count as an *elemental* analogue of graphene. Lithiation can potentially suppress the pseudo-Jahn-Teller distortion and thus flatten the silicene structure while opening the band-gap. Edge Si-Si bonds are shorter than inner ones of identical orientation, thus suggesting a contraction relaxation of the edge silicon atoms. An edge state appears at the Fermi level in broader zig-zag nanoribbons, but does not appear in all armchair nanoribbons due to their dimer Si-Si bond at the edge. All types of Si-Si bond display typical covalent bonding features, but their strength depends upon both the bond-orientation and the bond position. The larger the deviation of the Si-Si bond orientation from the nanoribbon axis, and the closer the Si-Si bond to the nanoribbon edge, the stronger is the strength of the Si-Si bond.[50] First-principles calculations led to the identification of so-called Janus silicene; silicene which is asymmetrically functionalized using hydrogen and halogen atoms. As compared with silicane, any Janus silicene system is a direct band-gap semiconductor. The band-gap of Janus silicene can have any value between 1.91 and 2.66eV. Elastic straining can further reduce the band gap to 1.11eV at a biaxial tensile strain of up to 10%.[51]

Mechanical Properties

Molecular dynamics simulations were used to investigate[52] the fracture of a silicene nanosheet, containing a pre-existing crack, under tensile loading at temperatures of 0, 100, 200, 400 or 500K. The crack angle was 0, 30, 45 or 90°. The ultimate tensile stress and associated strain decreased with increasing temperature. Nanosheet which was stretched at 300K, and contained a 1.5nm crack at an angle of 0°, gradually fractured with increasing strain, under increasing stress. A five-seven ring or a small cavity was observed when the strain was 0.2. When the strain was 0.25 to 0.3, the five-seven ring gradually became a small hole and propagated in a straight line at 45°. When the strain was 0.4, the sheet exhibited inward contraction at the edges. When the strain was 0.5, the sheet began to tear from the edges to the centre. When the crack angle was 90° and the tensile strain reached 0.225, the original line-like crack became an elliptical cavity. The latter enlarged as the strain increased. When the strain was 0.4, fracture spread from the sheet centre to the edge. When the angle was 30°, a five-seven-ring was observed when the strain was 0.2. An X-shaped crack was observed when the strain was 0.325. The fracture behaviour was different for angles of 0 and 90° because a shear component was introduced into the sheet and shear fracture was observed. When the crack length was increased, the strength of the nanosheet decreased and it broke easily. In every case, the nanosheet exhibited 5-7-rings and formed a small cavity which increased in size as the strain increased. When the strain was sufficiently large, an X-shaped crack always appeared because stress became concentrated near to the crack tips. A greater crack length reduced the ultimate stress for every crack-angle because a greater crack length introduced more localized stresses around the crack tip. A lower strain-rate led to a lower fracture strength because it provided more time within which to overcome the energy barrier. A longer crack then made the silicene nanosheet less sensitive to the strain-rate. The material had its most stable structure at 0K and five-seven rings were observed when the strain 0.175. These changed to five-eight rings when the strain was 0.325. No cavity was found, and an X-shaped crack was observed. The structural stability decreased as the temperature increased. At 100 to 500K, the structure changed from five-seven rings to the configuration observed at 0K and a cavity finally formed when sufficient strain was present. The behaviour was different to that observed at 0K because the strength gradually decreased as the temperature increased. At 500K, several large cavities formed at the centre of the sheet when the strain was 0.3. Those cavities became larger when the strain was 0.4. Higher temperatures led to earlier fracture of the sheet. At very low temperatures the silicon atoms underwent no thermal vibration and propagation of the crack tips was limited. At high temperatures, the silicon atoms possessed sufficient energy to overcome the activation energy barrier and several cavities could form. At 0, 100, 200, 400 and 500K, the ultimate stresses for nanosheets with a crack-length of 3nm and a crack-angle of 0° were 10.2, 9.8, 8.8, 7.9 and 7.4N/m, respectively. The respective strains were 0.35, 0.32, 0.28, 0.23 and 0.22.

Auxetic materials, which are 2-dimensional and have a negative Poisson ratio, exhibit interesting mechanical properties. First-principles swarm-intelligence structural-search methods were able to identify a thermodynamically-stable Si_2S monolayer[53]. This had a puckered 2-dimensional lattice in which sulphur atoms were adsorbed atop a distorted tetragonal silicene layer. The puckered lattice caused the Si_2S monolayer to exhibit in-plane Poisson ratios of -0.05 and -0.069 along the x and y directions, respectively. Electronic structure calculations showed that the Si_2S monolayer was a semiconductor with a quasi-direct band-gap of 1.81eV. This could be converted into a direct-gap semiconductor of 1.43eV by applying a tensile strain of about 2%. The monolayer had a visible-light absorption coefficient of 10^5/cm. The electron mobility was 81cm^2/Vs along the y direction. This was 1.5 times that along the x direction.

The ultimate tensile stresses which were predicted by atomistic reactive molecular dynamics simulations of monolayer silicene (tables 1 and 2) were analysed[54] in terms of the Weibull probability-of-failure method. Vacancy and interstitial defects were randomly distributed over a finite sheet of silicene which was subjected to tension in the zig-zag and armchair directions. This led to a marked decrease in strength, and there was a clear directional dependence: the mechanical properties in the zig-zag direction were the more greatly affected, for either defect-type. A single vacancy reduced the in-plane ultimate stress by some 13%, and a single interstitial or adatom led to a strength reduction of the order of 24% with respect to that of the pristine material. Statistical analysis of the tensile strengths indicated a Weibull modulus which was of the order of 50 to 150, depending upon the defect-type and the loading direction. Further analysis indicated a relatively high flaw tolerance even when defects were present.

Table 1. Effect of vacancy defects on the average ultimate in-plane stress, in-plane stiffness and ultimate strain of silicene

Number of Vacancies	Direction	Stress(N/m)	Stiffness(N/m)	Strain
0	zig-zag	5.85	50.44	0.1797
0	armchair	4.78	62.31	0.0886
1	zig-zag	5.11	48.49	0.1624
1	armchair	4.77	62.19	0.0883
3	zig-zag	4.79	44.69	0.1540
3	armchair	4.55	57.68	0.0953
5	zig-zag	4.37	40.73	0.1408
5	armchair	4.41	56.36	0.0914
8	zig-zag	4.00	43.71	0.1154
8	armchair	4.21	58.26	0.0997
10	zig-zag	4.18	38.52	0.1296
10	armchair	4.06	50.41	0.0930
13	zig-zag	3.95	38.69	0.1306
13	armchair	3.86	54.44	0.0987
16	zig-zag	3.79	34.79	0.1316
16	armchair	3.69	49.16	0.0885
18	zig-zag	3.88	37.76	0.1352
18	armchair	3.44	43.51	0.0846
20	zig-zag	3.34	29.26	0.1050
20	armchair	3.46	40.93	0.0849

Table 2. Effect of inclusion defects on the average ultimate in-plane stress, in-plane stiffness and ultimate strain of silicene

Number of Inclusions	Direction	Stress(N/m)	Stiffness(N/m)	Strain
0	zig-zag	5.85	50.44	0.1797
0	armchair	4.78	62.31	0.0886
1	zig-zag	4.45	45.05	0.1286
1	armchair	4.57	53.81	0.0883
3	zig-zag	4.30	42.46	0.1309
3	armchair	4.60	63.10	0.0894
5	zig-zag	4.20	47.13	0.0857
5	armchair	4.46	58.69	0.0845
8	zig-zag	4.38	41.99	0.1447
8	armchair	3.64	54.45	0.0791
10	zig-zag	4.22	44.56	0.0875
10	armchair	4.79	57.29	0.0477
13	zig-zag	3.86	43.70	0.1181
13	armchair	4.03	57.18	0.0856
16	zig-zag	3.91	41.06	0.1546
16	armchair	4.24	54.23	0.0894
18	zig-zag	3.66	43.25	0.1144
18	armchair	4.33	50.85	0.0992
20	zig-zag	3.69	41.38	0.1211
20	armchair	3.96	61.70	0.0368

Density functional theory simulations were made of the mechanical testing of silicene, and the Tersoff-potential parameters were optimised by using a gradient-based minimization method to match the density functional theory calculations in the elastic regime[55]. By using this parameter-set, the elastic moduli of silicene in the zig-zag and armchair directions were computed via molecular statics simulations and were compared with those of other silicon interatomic potential models and density functional theory results. The results of uniaxial tensile tests suggested that the angle penalty-function in the Tersoff model should be modified. Such a modification then led to improved stress-strain predictions.

The mechanical properties of silicene were investigated by means of molecular dynamics simulation[56]. Stress-strain analysis was performed for tensile loading, and the Young's modulus, tensile strength and fracture strain were deduced (table 3). The tensile strength of silicene sheet was lower than that of monolayer and bilayer graphene. In hydrogen-functionalized silicene, due to repulsion between the hydrogen atoms, those on both sides of the silicene sheet bonded alternately with silicon atoms. With increasing percentage of hydrogen-coating, from a single atom until complete coverage, the tensile strength and fracture strain decreased. It was noted that the degradation of the mechanical properties of functionalized graphene sheet was much greater than that of functionalized silicene sheet. The degradation of the mechanical properties became much more marked upon increasing the percentage of hydrogen-coating for functionalized graphene, as compared compared with that of functionalized silicene. The difference in behaviour was attributed to the fact that functionalization changed the hybridization of the carbon atoms in graphene from sp^2 to sp^3. Because of these change in hybridization, the graphene sheet lost its flatness so that, under tensile loading, the bonds of sp^3 could rotate. This caused the bonds to break. Silicon atoms in the silicene were alternately located out of the sheet due their greater atomic radius, and had an intrinsic sp^3 hybridization. Upon functionalization, the hybridization did not change in silicene sheet and the silicon atoms relocated only slightly because of the new bonds with hydrogen atoms. The functionalization of silicene therefore had a smaller effect upon its mechanical properties. The properties of functionalized silicene were examined for sheets with a linear hydrogen coating. Two patterns were considered: one in which the linear coating of hydrogen was perpendicular to the tension and one in which the linear coating was parallel to the tension. Simulations indicated that the tensile strength and fracture strain for a linear pattern perpendicular to the tension decreased more than when the coating was parallel to the loading direction. In silicene sheet where the linear pattern was perpendicular to the loading direction, the hydrogen atoms bonded with silicon atoms from both sides of the sheet. When the linear pattern was parallel to the loading direction, the hydrogen atoms

Materials Research Forum LLC

https://doi.org/10.21741/9781644903650

bonded with silicon atoms from one side of the silicene sheet. The silicon atoms were therefore pulled more out of the silicene sheet, when the pattern was perpendicular to the loading direction, than in the other case. The weakening perpendicular line then had a greater effect upon the strength of silicene sheet.

Table 3. Tensile strength and fracture strain of silicene

Type	Tensile Strength(GPa)	Fracture Strain
armchair	120	0.2
zig-zag	134	0.18

The structural and mechanical properties of silicene were investigated by means of density functional theory calculations. In order to calculate the Young's modulus, the bulk modulus, the shear modulus and the Poisson ratio, unit cells which contained two atoms were used and the effect of chirality upon the elastic properties was determined[57]. The silicene exhibited isotropic behaviour, unlike the anisotropic behaviour of graphene. The calculated moduli for silicene were markedly lower than those of graphene in the zig-zag and armchair directions, while the Poisson ratio of silicene was higher than that of graphene. Non-harmonic interatomic potentials were used to construct non-linear models for the material by using a modified Morse potential function. In this way, the effects of chirality on dissociation energy, inflection point and coefficients of the modified Morse potential function were deduced. A comparison of the cut-off distance in the modified Morse potential showed that the inflection-point values for armchair and zig-zag graphene were highly direction-dependent. The values were hardly different for silicene.

Finite-element methods were used to investigate the fracture of a silicene nanosheet[58]. The latter was treated as a buckled framework and the mechanical behaviour was simulated by using a modified Morse potential function. The interatomic force per atom was calculated as if part of a set of non-harmonic oscillator networks; described by the modified Morse potential. The non-linear behaviour with increasing strain was defined by those interatomic forces, using an iterative method. Non-linear stress-strain relationships for armchair and zig-zag silicene nanosheets were obtained for pristine and defective material (table 4). For both the armchair and zig-zag configurations, there was a sudden drop in the stress-strain diagram, reflecting a brittle behaviour. The tensile strength and ultimate strain of armchair silicene were deduced to be slightly larger than those of zig-zag silicene. The mechanical properties were markedly affected by single-vacancy and Stone-Wales defects. A single-vacancy defect could reduce the ultimate strain by about 7.3%. The effect of a Stone-Wales defect was less marked. The tensile strengths of the armchair and zig-zag nanosheets were 12.87 and 12.31GPa, respectively. The ultimate strains under the same conditions were 38.16 and 34.14%.

Table 4. Tensile strength and ultimate strain of silicene nanosheets

Defect	Material	E(TPa)	Tensile Strength(GPa)	Ultimate Strain(%)
none	armchair	0.0618	12.87	38.16
Stone-Wales	armchair	0.0604	12.80	36.72
single vacancy	armchair	0.0614	12.18	35.38
1% defect	armchair	0.0604	11.28	32.06
2% defect	armchair	0.0566	10.01	30.53
3% defect	armchair	0.0509	7.98	27.90
none	zig-zag	0.0546	12.31	34.14
Stone-Wales	zig-zag	0.0548	12.17	33.30
single vacancy	zig-zag	0.0538	12.04	32.50
1% defect	zig-zag	0.0531	10.54	29.61
2% defect	zig-zag	0.0502	9.01	28.02
3% defect	zig-zag	0.0455	7.95	25.39

The properties of silicene under tensile, bending and oscillation conditions were simulated by using molecular dynamics techniques[59]. Plastic flow occurred, and an elongation of 46% was observed in a 200Å x 700Å sheet at 298K during tensile testing. The yield strength, ultimate tensile strength, Young's modulus, cohesive energy and bulk modulus were found to be 18.28GPa, 23.96GPa, 5.25TPa, 3.72eV/atom and 3.62TPa, respectively. The Poisson ratio was 0.75. A strength greater than 0.178TPa was observed.

The mechanical properties and failure of a graphene|silicene|graphene heterostructure were investigated by using molecular dynamics simulations[60]. By sandwiching silicene between the graphene layers, the ultimate tensile strength and Young's modulus of the heterostructure were increased by a factor of about 10 over those of standalone silicene.

The graphene and silicene exhibited entirely different fracture behaviours: the graphene cleaved only through its zig-zag edge while silicene could cleave through its zig-zag and armchair edges. An increase in temperature led to a decrease in its strength and stiffness, whereas an increase in strain-rate led to an increase in strength with no great change in stiffness. The temperature and strain-rate dependent fracture stresses could be accurately described by the kinetic theory of fracture.

Molecular dynamics methods were used to simulate the tensile testing of multi-layered silicene, and the fracture and elastic properties of one- to five-layered armchair and zig-zag silicenes were determined[61]. The fracture stress of multi-layered armchair silicene was greater than that of zig-zag silicene having the same number of layers. The fracture properties of the multi-layered armchair and zig-zag silicenes were inversely related to the number of layers. The elastic modulus of the multi-layered material did not exhibit much variation as a function of nanosheet size. The fracture of the armchair and silicene nanosheets began with the appearance of vacancy and Stone-Wales defects within the structure of the nanosheets.

Silicane, fully hydrogenated silicene, was studied with regard to tailoring its electronic properties via straining[62]. The mechanical response to strain was predicted by performing first-principles calculations based upon density functional theory. The elastic limits were 0.22, 0.28 and 0.25 for the armchair, zig-zag and biaxial directions, respectively. These were increases of 29, 33 and 24%, respectively, over those of silicene. The in-plane stiffness and Poisson ratio magnitudes were smaller by factor of 16 and 26%, respectively. On the other hand, hydrogenation and dehydrogenation had little effect upon the ultimate tensile strength. The highest-order elastic constants (table 5) were measured in order to furnish the best continuum description of the non-linear elastic response. The second-, third-, fourth- and fifth-order elastic constants were within the strain-range of 0.02, 0.08, 0.13 and 0.21, respectively. The pressure-effect upon the second-order elastic constants and Poisson ratio was deduced from the third-order elastic constants.

Table 5. High-order elastic constants of silicene

Material	Constant	Value
silicane	C_{11}	57.1
silicene	C_{11}	71.3
silicane	C_{12}	13.7
silicene	C_{12}	23.2
silicane	C_{111}	-404.9
silicene	C_{111}	-397.6
silicane	C_{112}	-34.6
silicene	C_{112}	-14.1
silicane	C_{222}	-349.6
silicene	C_{222}	-318.9
silicane	C_{1111}	2146
silicene	C_{1111}	-830
silicane	C_{1112}	-195
silicene	C_{1112}	-309
silicane	C_{1122}	-107
silicene	C_{1122}	-5091
silicane	C_{2222}	1240
silicene	C_{2222}	-629
silicane	C_{11111}	-7525
silicene	C_{11111}	-20614
silicane	C_{11112}	2495
silicene	C_{11112}	6923
silicane	C_{11122}	-1671
silicene	C_{11122}	11681

silicane	C_{12222}	-3318
silicene	C_{12222}	-7593
silicane	C_{22222}	-4517
silicene	C_{22222}	-29735

First-principles theoretical calculations were made of the mechanical properties of the silicene and silicane structures (tables 6 and 7) under uniaxial tensile strain[63]. Chirality effects were more appreciable in the case of silicene than that of silicane. A band-gap of up to 0.8eV developed in silicene under uniaxial tension, and vanished before it attained its in-plane ultimate strength. On the other hand, a pre-existing band-gap of 2.39eV existed in the silicane, and decreased to zero with increasing tensile strain without chirality effects. Failure in the silicene structure began with instability of the slightly buckled lattice structure. In silicane, the structure failed due to weakened Si–Si bonding. Under tensile testing in the armchair direction, an indirect-direct gap transition occurred at a strain of 5%.

Table 6. Comparison of silicene and silicane structures

Material	Si-Si Bond(Å)	Bond Angle(°)	Buckling(Å)	Si-H Bond(Å)
silicene	2.28	116.25	0.45	-
silicane	2.36	111.09	0.72	1.50

Table 7. Comparison of silicene and silicane properties

Material	Chirality	In-Plane Stiffness(N/m)	Poisson Ratio	Ultimate Stress(N/m)
silicene	zig-zag	60.06	0.41	5.66
silicene	armchair	63.51	0.37	7.07
silicane	zig-zag	54.50	0.25	7.93
silicane	armchair	54.79	0.23	7.29

An in-plane stiffness of 52.55N/m and a Poisson ratio of 0.24 were calculated[64] for silicane by calculating the strain energy. When compared with silicene, the silicane was softer due to its relatively weaker Si-Si bonds. The band-structure of silicane could be tuned by applying a uniform tensile strain; an increase in which monotonically decreased the band-gap. The silicane also underwent an indirect-to-direct gap transition under small strains, and a semiconductor-to-metal transition under large strains. An electric field could markedly change the Si-H bond-length. When a strong field was applied, the hydrogen atom on the high-potential side became desorbed, while the hydrogen atom on the low-potential side remained bonded. An external electric field could therefore create single-sided hydrogenated silicene from silicane.

Molecular dynamics methods were used to simulate the deformation of silicene when subjected to a tensile load which was applied parallel to the armchair direction[65]. In order to explore the effect of specimen size, models which comprised 600, 20000, 60000 or 100000 atoms were used. Periodic boundary conditions were imposed in all three directions, and a time-step of 0.002ps was used.The simulations assumed a temperature of 50K, and the strain-rate was 10^9, 10^{10} or 10^{11}/s (table 8). The Young's modulus was independent of strain-rate for a model specimen comprising 100000 atoms, while the fracture strain depended upon the specimen size and the strain-rate. The Young's modulus decreased at high strain-rates. The ultimate tensile strength, fracture stress and fracture strain of the nanosheets increased with increasing strain-rate. sheet is observed with increasing strain rate. The structure became distorted during the initial stages of deformation, but reverted to its original planar hexagonal form at higher strains. The material exhibited a high ductility at higher strain rates, due to the initiation of multiple cracks which produced more granules which were connected by threads.

Materials Research Forum LLC

https://doi.org/10.21741/9781644903650

Table 8. Properties of simulated silicene specimens

Atoms	Strain-Rate(/s)	UTS(GPa)	Fracture Stress(GPa)	Fracture Strain
600	10^9	35.7	9.02	0.292
600	10^{10}	56.9	34.2	0.312
600	10^{11}	63.83	37.67	0.460
20000	10^9	35.09	10.1	0.296
20000	10^{10}	56.61	27.5	0.706
60000	10^9	34.62	9.68	0.300
60000	10^{10}	56.37	27.42	0.890
100000	10^9	34.88	6.17	0.662
100000	10^{10}	56.53	27.47	0.840

The effects of vacancies and Stone-Wales defects upon the mechanical properties of silicene were investigated by using molecular dynamics finite-element methods and the Tersoff potential[66]. The Young's modulus, Poisson ratio and uniaxial tensile stress-strain behaviour (table 9) were studied for the armchair and zig-zag directions. Pristine and slightly-defective sheets exhibited almost the same elastic behaviours up to fracture. A single defect could greatly weaken the sheet, leading to a marked reduction in the fracture strength. A single 2-atom vacancy at the sheet-centre reduced the fracture stress by 18 to 20 % and the fracture strain by 33 to 35%. The weakening effect of the Stone-Wales defect depended upon the tensile direction and the orientation of the defect. For a given axial strain, the uniaxial stress in the zig-zag direction was slightly higher than that in the armchair direction. The 2-dimensional Young's modulus of pristine sheet was estimated to be about 60N/m. The Poisson ratio in the zig-zag direction was always higher than that in the armchair direction. The fracture stress (8.2N/m) and fracture strain (23.2%) of the pristine sheet in the zig-zag direction were slightly higher than those (7.9N/m and 21.2%) for the armchair direction. For a given axial strain before fracture, the Poisson ratio of the defective sheet was almost equal to that of the pristine sheet. A Stone-Wales defect affected both the fracture strength and the failure mode of the defective sheet.

Table 9. Tensile properties of silicene sheets

Direction	Defect	E(N/m)	Maximum Stress(N/m)	Strain(%)
zig-zag	none	60.0	8.23	23.2
zig-zag	single 2-atom vacancy	59.7	6.74	15.0
zig-zag	Stone-Wales 1	59.7	6.85	15.2
zig-zag	Stone-Wales 2	59.9	6.61	14.3
armchair	none	59.7	7.95	21.2
armchair	single 2-atom vacancy	59.6	6.35	14.1
armchair	Stone-Wales 1	59.6	6.79	15.8
armchair	Stone-Wales 2	59.5	7.51	18.9

Molecular dynamics simulations were used to determine the deformation mechanism of 2-dimensional metamaterials[67]. These revealed that deformation under axial tensile loading could be divided into two stages: bending-dominated and stretching-dominated. The rotation of junctions in the first stage resulted in an auxetic behaviour of the metamaterial. The auxetic behaviour depended greatly upon the amplitude/wavelength ratio (figure 2) of the sinusoidal ligaments of the simulation and upon the stiffness ratio between the axial and transverse directions. The magnitude of the negative Poisson ratio increased from 0 to 0.625, with a related increase in the amplitude/wavelength ratio from 0 to 0.225. The magnitude of the negative Poisson ratio increased from 0.47 to 0.87 as the stiffness ratio increased from 0.125 to 8. Due to the anisotropy of the underlying material, the Poisson ratio of a proposed 2-dimensional lattice structure depended greatly upon the stiffness ratio of the axial group and the transverse group. The stiffness ratio was defined as being the ratio of Young's moduli in the x and y directions. It could be changed by varying the bond stiffness in the y direction, while the bond stiffness and Young's modulus in the x direction remained constant. There were marked differences between the Poisson ratios of samples with differing stiffness ratios. The Poisson ratio increased as the stiffness ratio increased. The stiffness ratio was the predominant factor determining the Poisson ratio of a 2-dimensional lattice structure. In some cases, there was expected to be a sharp change in Poisson ratio from negative to positive; indicating out-of-plane buckling. In other cases, there was expected to be a smooth transition of the Poisson ratio from negative to zero. Out-of-plane buckling was attributed to a competition between the out-of-plane bending and in-plane bending of transverse groups. When close to a critical point, the considerable bending-energy stored in the transverse group induce out-of-plane buckling. The bond- and angle-energies, which were indicators of in-plane bending, underwent a sharp drop while the dihedral energy, which was an indicator of out-of-plane buckling, underwent a sharp increase. The critical strain depended markedly upon the ratio between out-of-plane bending stiffness and in-plane bending stiffness. In the absence of out-of-plane buckling, the Poisson ratio would exhibit a smooth transition from negative to zero. The initial negative Poisson ratio resulted from a so-called straightening and elongation of the transverse group which was caused by rotation of junctions due to bending-deformation of the axial group. When the axial group fully straightened, deformation of the axial group could not cause the rotation of junctions; leading to the zero Poisson ratio. Because the out-of-plane bending stiffness was large enough, 2-dimensional lattice structures could exhibit a smooth transition from negative to zero Poisson ratio.

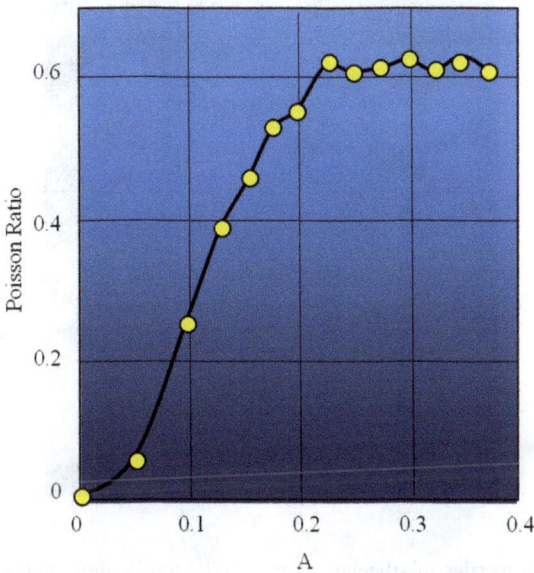

Figure 2. Poisson ratio versus amplitude/wavelength ratio

Mode-I stress intensity factors were estimated[68,69] and the crack-tip displacement field which was predicted by linear elastic fracture mechanics was used to determine the stress intensity factor (table 10). Atomic-scale finite-element methods and Stillinger-Weber potentials were used to simulate tensile tests. An empirical formula was derived which related the mode-I fracture toughness to the elastic modulus, the intrinsic tensile strength, the bond-length and the buckling height of atom-thick hexagonal material.

Table 10. Properties of silicene

Property	Value
lattice constant	3.870Å
bond length	2.279Å
buckling height	0.449Å
Van der Waals radius	2.1Å
E(armchair)	62.2N/m
E(zig-zag)	62.6N/m
Poisson ratio(armchair)	0.15
Poisson ratio(zig-zag)	0.15
maximum stress(armchair)	8.0N/m
maximum stress(zig-zag)	7.0N/m
strain(armchair)	20.75%
strain(zig-zag)	20%
K_I(armchair)	0.70MPa√m
K_I(zig-zag)	0.72MPa√m

The mechanical properties of silicene under uniaxial tensile deformation were modelled by using molecular dynamics simulations. The fracture strength and fracture strain were much higher than those of bulk silicon, but the Young's modulus of silicene was lower than that of bulk silicon. Increasing the temperature markedly decreased the fracture strength and fracture strain of silicene, while increasing the strain-rate slightly increased them[70]. Brittle fracture was also observed. Stress-strain curves showed that no apparent plastic deformation occurred before fracture. The stress fell suddenly upon attaining the maximum strength. The silicene thus exhibited brittle fracture under tension deformation. Brittle failure was also found during simulated fracture. Although tension deformation was applied uniformly over the whole area of the silicene, the atomic stresses were not uniformly distributed, due to thermal fluctuations at a finite temperature. At a strain of 0.182, a small crack nucleated via the breaking of a Si-Si bond when it reached the critical bond-length. The crack then spontaneously propagated, leading to cleavage failure without plasticity.

Molecular dynamics simulations were used[71] to investigate the mechanical properties of polycrystalline silicene. Annealing was used to construct a more realistic model for the material. A more stable structure was formed due to the breaking and reforming of bonds between atoms on the grain boundaries. As the grain-size decreased, the efficiency of the annealing process increased. Biaxial tensile tests were performed on annealed samples in order to determine the relationship between the grain-size and properties such as the in-plane stiffness, fracture strength and fracture strain (figures 3 and 4). As the grain-size decreased, the fracture strain increased while the fracture strength exhibited an inverse tendency. The decreased fracture strength was partly attributed to a weakening effect which resulted from the increasing areal density of defects which acted as a reservoir of stress-concentrated sites on the grain boundary.

Molecular dynamics simulations of hydrogen-functionalized silicene nanosheets (silicane) show that the mechanical properties of the nanosheets are degraded by the functionalizing. Upon comparing the mechanical properties of armchair and zig-zag silicene and silicane nanosheets, it is found that armchair nanosheets have a higher Young's modulus, fracture strength and fracture strain than do zig-zag silicene and silicane nanosheets of the same dimension. As for the fracture pattern of silicane nanosheets, brittle behaviour is observed for both armchair and zig-zag forms.[72] The mechanical properties of both armchair and zig-zag silicane exhibit an inverse dependence upon temperature.

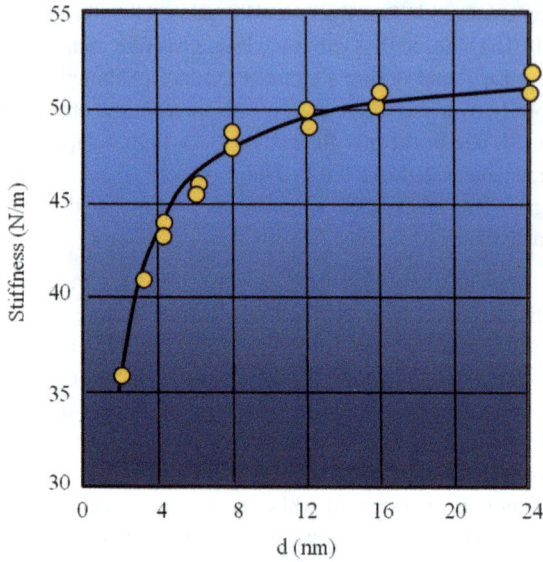

Figure 3. Grain-size dependence of the in-plane stiffness of silicene

The mechanical properties of silicene nanostructures subject to tensile loading have been studied via molecular dynamics simulation. The maximum in-plane stress and the corresponding critical strain of the armchair and the zig-zag silicene sheets at 300K were 8.85 and 10.62, and 0.187 and 0.244N/m, respectively. The in-plane stresses of the silicene sheet in the armchair direction at 300, 400, 500 and 600K were 8.85, 8.50, 8.26 and 7.79N/m, respectively. The in-plane stresses of the silicene sheet in the zig-zag direction at 300, 400, 500 and 600K were 10.62, 9.92, 9.64 and 9.27N/m, respectively.[73] The silicene sheet yielded in the zig-zag direction as compared with tensile loading in the armchair direction. Wrinklons and waves were observed at the shear band across the centre-zone of the silicene sheet.

Materials Research Forum LLC
https://doi.org/10.21741/9781644903650

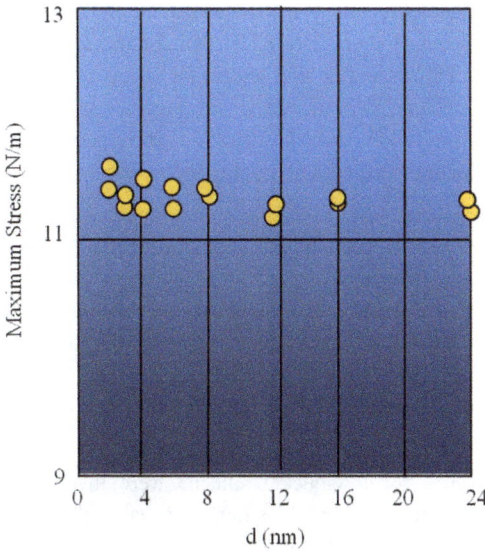

Figure 4. Grain-size dependence of the ultimate strength of silicene

The influence of defects upon the mechanical properties and failure behaviour of silicene sheets, when investigated using molecular mechanics and molecular dynamics methods, is such that the intrinsic strength of the sheets decreases with increasing linear density of vacancies, increasing width ratio of cracks and increasing inflection angle of grain boundaries. The elastic properties of sheets are affected not only by defects but also by their corrugated structure. The fracture failure of sheets with defects usually starts from Si-Si bonds located at the defect edge.[74] The stretching meanwhile tunes the electronic structure of the sheets.

Analytical expressions for the central, k_0, and non-central, k_1, force constants of two-dimensional (graphene, silicene) and three-dimensional (diamond, silicon) structures have been obtained within the previously proposed model of the binding energy of carbon atoms in graphene. The Kleinman internal displacement parameter of the two-dimensional structure has been determined and this shows that the ratio, k_0/k_1, depends only upon the dimension of the structure.[75]

Fully atomistic first-principles molecular dynamics methods predict an elastic stiffness of 50.44N/m for the zig-zag direction and 62.31N/m for the armchair direction; with an ultimate strength of 5.85N/m and an ultimate strain of the order of 18% for monolayer silicene. A weak directional dependence is observed. A predicted effective bending stiffness of 38.63eV/unit-width indicates that its corrugated structure increases the bending rigidity, compared to the similar graphene system[76].

Density functional based tight-binding and reactive molecular dynamics studies of the mechanical properties of suspended single-layer silicene yield Young's modulus values for armchair and zig-zag membranes upon fitting the linear sections of predicted stress-strain curves. Very small differences are found between the values for membranes having differing edge terminations. For armchair membranes, values of 43N/m (0.043TPa nm) and 62.7N/m (0.0627TPanm) are found using various techniques. For zig-zag membranes, values of 43N/m (0.43TPanm) and 63.4N/m (0.0634TPanm) are found, depending upon the technique[77].

Stress-strain relationships for low-buckle silicene were investigated[78], using density functional theory, for equiaxial tensile strain and uniaxial tensile strain along the armchair and zig-zag directions. The ideal strengths for equiaxial tension and armchair uniaxial tension were 7.59 and 6.76N/m, respectively. In the case of zig-zag uniaxial tension, there were ideal strengths of 5.26 and 5.29N/m, due to a phase transition from the original low-buckle structure to a truly planar structure. The phonon dispersion under three types of tension was calculated by using finite-difference methods, showing that the failure mechanism under armchair and zig-zag uniaxial tension was elastic instability. Phonon instabilities occurred near to the centre of the Brillouin zone, and the phonon soft modes for armchair and zig-zag uniaxial tension were longitudinal acoustical modes along the tensile direction. The failure mechanism of silicene under equiaxial tension could be attributed only to elastic instability; contrary to the case of graphene. The phonon instability was governed by the out-of-plane acoustical mode rather than the K1 mode in graphene. Phonon instability also occurred in the planar structure of silicene which transformed under zig-zag uniaxial tension. Uniaxial tension along the armchair and zig-zag directions could not open a gap in silicene but could vary the gaps which were introduced by spin-orbital coupling by applying the two types of tension.

Density functional theory analysis of the mechanical stabilities of planar and low-buckled honeycomb monolayer structures of silicon under large strains showed that both types could sustain large strains (≥ 0.15) under armchair, zig-zag and biaxial deformation. Knowledge of the third-, fourth- and fifth-order elastic constants was essential for accurate modelling of mechanical properties under strains greater than 0.03, 0.06 and

Materials Research Forum LLC
https://doi.org/10.21741/9781644903650

0.08, respectively. The second-order elastic constants, including the in-plane stiffness, were predicted to increase monotonically with pressure while the Poisson ratio monotonically decreased with increasing pressure. Results on the positive ultimate strengths and strains, second-order elastic constants and in-plane Young's modulus indicated that both forms of silicon were mechanically stable[79].

First-principles calculations predicted that the in-plane stiffness of silicene was much lower than that of graphene. The ideal yield strain of silicene under uniform expansion was about 20%. Homogeneous strain could introduce a semimetal-to-metal transition. The semimetallic state of silicene, in which the Dirac cone was located at the Fermi level, could persist only up to tensile strains of 7%. At larger strains, silicene changes into a conventional metal. The armchair Young's modulus was 63.00GPa.nm, while the zig-zag Young's modulus was 51.00GPa.nm.[80] Others had estimated the armchair Young's modulus to be 61.33GPa.nm.[81]

First-principles density functional theory calculations were used to investigate the mechanical properties of single-layer and free-standing silicene. Uniaxial tensile and compressive simulations were used to deduce the Young's modulus, Poisson ratio and ultimate strength (table 11). Loading in the armchair direction led to higher elastic moduli and tensile strengths, but to a lower Poisson ratio and smaller ultimate tensile elongation, than those for loading in the zig-zag direction. It was predicted that armchair loading led to the elongation of all bonds whereas zig-zag loading affected only the length of bonds which were partially oriented along the loading direction. Loading might not alter the lengths of bonds which were perpendicular to the tensile direction. For all of the structures the buckling increased under compressive loading, while it decreased under tensile loading. This tendency was very marked for extension along the zig-zag direction at higher loadings.

Table 11. Properties of silicene sheets, predicted by density functional theory

Direction	Young's Modulus(GPa)	Poisson Ratio	UTS(GPa)
zig-zag	59	0.33	6.0
Armchair	61.7	0.29	7.2

Ab initio density functional theory calculations of the properties of multi-layers with various stacking configurations revealed the evolution of those properties as the number, n, of layers ranged from 1 to 10. Strong interlayer covalent bonding existed between the layers in multilayers of silicene, unlike the weak van der Waals bonding between graphene layers. The interlayer bonding strongly affected the geometrical and electronic structures of the multilayers. As in the case of graphene bilayers, silicene with two different stacking configurations, AA and AB, exhibited linear and parabolic dispersions, respectively, around the Fermi level. When n was greater than 3, the geometrical and electronic properties of multilayers with 4 different stacking configurations, AAAA, AABB, ABAB and ABC, could be considered. The 3 stacking configurations, AAAA, AABB and ABC, which have a tetrahedral coordination, exhibited higher cohesive energies than did the Bernal, ABAB, stacking configuration. This again differed from the case for multilayers of graphene, where ABAB is the lowest-energy configuration.

Molecular dynamic simulations of silicene nanoribbons of various widths, under 5 or 10% uniaxial strain at 1 and 300K, predicted that they would be very ductile, would possess considerable toughness and would exhibit a very long plastic range before fracture. Under uniaxial strain, the nanoribbon structure also gradually changed from 2-dimensional to 1-dimensional[82].

Molecular dynamics simulations, in canonical ensemble, of large deformations of 2-dimensional silicene nanosheet under uniaxial and biaxial tension predicted that the Young's and bulk moduli, and ultimate tensile stress, were lower than those of graphene. The ultimate strain was higher than that of graphene for armchair silicene, unlike the zig-zag form. The Poisson ratio of silicene was also predicted to be greater than that of the carbon counterpart; due to the longer Si-Si bond-lengths and the low-buckle honeycomb structure. The bulk modulus was strongly size-dependent and decreased with increasing length of the nanosheet. Under large deformations, the formation of topological defects and silicon chains was predicted. Silicene is noticeably weaker than graphene in the zig-zag direction[83].

Molecular dynamics simulation, using an environment-dependent interatomic potential, was used to describe the interaction of silicon atoms and to predict the Young's modulus of defect-free and defected silicene nanoribbons. This showed that the moduli of pristine and defected nanoribbons increased with ribbon-length in both chirality directions. The Young's modulus of defected nanoribbons exhibited a complex dependence upon vacancies. With respect to temperature, the Young's modulus of nanoribbons, with and without vacancy defects, exhibited a non-linear behaviour which might possibly be tailored for a given length and chirality[84].

When the mechanical properties of silicene were investigated using *ab initio* calculations and molecular dynamics simulations with various empirical potentials, the results showed that the Young's modulus for bulk silicene was consistent with the *ab initio* calculations. The chirality had a significant effect upon the critical strain and stress of bulk silicene under uniaxial tension. The Young's modulus also depended strongly upon the chirality and size of a nanoribbon, due to edge-effects[85].

Ab initio calculations of the effects of biaxial tensile strain predicted that, up to a strain of 5, the Dirac cone remains essentially at the Fermi level while higher strains induced hole-doped Dirac states because of weakened Si-Si bonding. The lattice was expected to be stable up to a strain of 17. Buckling decreased with strains of up to 10, and then increased again; accompanied by a band-gap variation. The strain-dependence was similar to that of graphene. Values of the Grüneisen parameter were also calculated (table 12).[86]

Table 12 Grüneisen parameter of silicene as a function of strain

Strain(%)	γ_G
5	1.64
10	1.62
15	1.54
20	1.34
25	1.42

First-principles density functional theory calculations predicted that buckled single-layer silicene could transform into planar hexagonal silicene at a critical tensile strain of 0.20. Phonon dispersion analysis suggests that planar hexagonal silicene was stable under tension. The Poisson ratio exhibited a marked anisotropy: it increased when stretched in the zig-zag direction, but decreased when strained in the armchair direction[87]. When stretched in the former direction, the Poisson ratio of silicene could attain 0.62.

A hierarchical first-principles investigation of the entangled effects of lattice dimensionality and bond characteristics in lattice dynamics showed that bond-bending contributed negatively to the Grüneisen constant, γ, resulting in a negative acoustic γ. Bond-stretching similarly contributed positively to the Grüneisen constant, resulting in a positive optical γ. Layer-thickening caused by chemical functionalization tended to increase the acoustic γ, due to the increased bond-stretching effect. Bond-weakening caused by chemical functionalization tended to decrease the optical γ, due to the decreased bond-stretching effect. Excitation of the negative-γ modes resulted in a negative thermal expansion, while mode excitation and thermal expansion competed thermomechanically.[88]

During the synthesis of silicene, structural defects such as edge cracks are likely to be generated and affect its properties. Molecular dynamics simulations of armchair nanoribbons with edge cracks showed that the mechanical properties were impaired by edge cracks. Pristine nanoribbons, and those with edge cracks, exhibited brittle fracture. The crack-length played an important role in determining the critical strain and the fracture strength. Investigations of the effect of strain-rate and temperature upon the mechanical properties of nanoribbon with edge cracks showed that an increasing strain-rate increased the critical strain and fracture strength, but decreased the Young's modulus[89]. Low strain-rates also changed the expanded directions of cracks. Increasing the temperature could also significantly impair the mechanical properties of cracked nanoribbons.

The presence of a substrate can be expected to have an effect upon film properties. First-principles calculations of competing silicene phases under in-plane compressive stress, when either free-standing or deposited onto ZrB_2 (00•1) surfaces, revealed a particular ($\sqrt{3}$ x $\sqrt{3}$)-reconstruction to be stable on the ZrB_2 under epitaxial conditions. Unlike planar and buckled forms of free-standing silicene, all but one of the silicon atoms per hexagon of this planar-like phase, resided in a single plane. Without a substrate, and for a wide range of strains, this phase was energetically less favorable than the buckled one. The atomic positions were determined by interactions with the nearest-neighbour zirconium atoms competing with Si-Si bonding interactions that were imposed by the constraints of the honeycomb lattice[90]. The preparation of silicene-free ZrB_2(00•1) thin films, grown onto Si(111) by Ar^+-ion bombardment, permitted study of the spontaneous formation of silicene on their surfaces. Imaging of the ZrB_2(00•1) surfaces using scanning tunnelling microscopy revealed the structures of Zr-terminated and B-terminated ZrB_2(00•1) created by the bombardment.[91] The spontaneous formation of a continuous silicene sheet on a sputtering-induced disordered ZrB_2 surface demonstrated that silicene did not require an atomically-flat crystalline template in order to be stabilized.

The structure and stability of lithium-adsorbed silicene under biaxial strain were analysed by using first-principles plane-wave pseudopotential methods based upon density functional theory. Lithium-adsorbed silicenes retained their original configurations when tensile, and some compressive, strains were applied. The silicene plane meanwhile bulged towards the lithium atom when a larger compressive strain was applied, and the overall energy of the system became much lower. Lithium-adsorbed silicene tended to be unstable under compression[92].

Going one stage further, actual reaction with gases could be expected to have an effect. First-principles calculations of fully-oxidized silicene predicted that the zig-zag ether-like conformation was the most energetically favorable structure. These nanosheets had interesting elastic characteristics and even exhibited an auxetic behaviour, with a negative Poisson ratio. Due to the anisotropic mechanical and electronic properties, the material possessed a high axial intrinsic charge-mobility of up to $10^4 cm^2/Vs$; comparable to that of graphene nanoribbons[93].

First-principles investigation of halogenated silicene, SiX (X = F, Cl, Br or I), predicted that such materials exhibited an enhanced stability as compared with silicene and possessed a tunable direct gap with a small carrier effective mass. The halogen-dependence of the energy gap could be understood in terms of variations in buckling and bond-energy perturbation based upon orbital hybridization. A negative Poisson ratio was predicted in the case of fluorinated silicene with a boat structure[94].

Although the present book limits itself to discussing elemental analogues, it is nevertheless interesting to see how the properties of silicene and graphene 'blend into' one another. The Harrison bonding-orbital method and the Keating model, when applied to the concentration dependences of the elastic constants of the two-dimensional Si_xC_{1-x} system, predict that the central and non-central force constants and the Grüneisen parameter all exhibit a non-linear behaviour over the transition from graphene to silicene. Short-range repulsion plays a non-trivial role in this. It is also found that the elastic constants and Young's modulus change almost linearly over the transition from graphene to silicene[95].

First-principles calculations of the atomic structures and non-linear properties of single-layer graphene, bilayer graphene, single layer silicene and bilayer silicene under equiaxial tension and uniaxial tensions along armchair and zig-zag directions show that there exists a weak Van der Waals interaction between the two layers of bilayer graphene, and that the interlayer distance is not variable with strain for the three types of tension. On the other hand, the interlayer of bilayer silicene involves a covalent bond interaction and the distance decreases with increasing strain for all three types of tension. A continuum description of the elastic response is achieved by expanding the elastic-strain energy-density as a Taylor series in strain; truncated after the third-order term. The in-plane second- and third-order elastic constants of bilayer graphene and bilayer silicene are obtained by fitting the strain energy density versus Lagrangian strain relationships. The results predict that the in-plane stiffnesses of bilayer graphene and bilayer silicene become slightly greater than those of their single-layer counterparts. In spite of the interlayer Si-Si covalent bond between two layers of bilayer silicene, its stiffness is still much lower than that of bilayer graphene and single-layer graphene. The Poisson ratios of bilayer graphene and bilayer silicene remain essentially unchanged as compared with their single-layer counterparts[96].

In parallel with the effect of various factors determining the mechanical properties of silicene, there is the reciprocal effect of straining upon other properties. Straining can be an effective means for continuously tuning the electronic and optical properties of materials[97]. The thermal conductivity of 2-dimensional materials typically decreases when a tensile force is applied, due to a softening of the phonon modes. Variations in the thermal conductivity of free-standing strained silicene were compared theoretically with that of unstrained silicene. The analysis was based upon the phonon Boltzmann transport equation, within the relaxation-time approximation, and used strain-dependent force-constants which were deduced from first-principles calculations. Of particular interest was a competition between the opposed effects of a strain-induced enhancement and that of vacancy defects which produced a reduction in thermal conductivity. The existence of the latter exerted a much stronger effect on the strained silicene, and was able to remove or move a peak in the conductivity of a theoretically infinite silicene sample. Such conductivity-suppression by vacancies depended upon the degree of strain. With increasing temperature, the conductivity became less sensitive to strain and any difference in the predictions for infinite and finite-sized samples gradually disappeared.

Non-equilibrium molecular dynamics simulations of thermal transport in single-layer sheet under uniaxial straining showed[98] that, unlike graphene, silicene exhibited an anomalous response to tensile force. The thermal conductivity of sheet and nanoribbons first increased markedly with strain, rather than decreasing, and then fluctuated around a higher plateau value. The phonon transport in silicene was dominated by out-of-plane flexural modes, as in graphene. The differing behaviour of silicene was attributed to an interplay between uniaxial stretching modulation in the longitudinal direction (which markedly decreased the phonon group-velocities of longitudinal and transverse modes and thus hindered heat conduction), and phonon-stiffening of the flexural modes (which counteracted the phonon-softening effect and aided thermal transport). The behaviour of the silicene sheet was also related to the deformation characteristics of the hexagonal lattice.

The effect of biaxial strain upon the thermo-electric transport coefficients of slightly-buckled monolayer silicene was determined[99] by using first-principles methods. Compressive strain led to lattice instability while tensile strain led to stability, with a small bandgap-opening at the K-point. With increasing tensile strain, the Seebeck coefficient and electronic thermal conductivity improved while the electrical conductivity and lattice thermal conductivity decreased slightly.

Upon applying uniaxial and biaxial strains, the band-gap opened at small strains but, upon increasing the strain beyond 8%, silicene changed into a metal[100]. The ultimate

Materials Research Forum LLC

https://doi.org/10.21741/9781644903650

tensile strength was 3.4GPa. The imaginary part of the dielectric function showed that inter-band transitions were red-shifted for uniaxial and biaxial tensile strains and were blue-shifted for uniaxial and biaxial compressive strains. Electron energy-loss data showed that the $\pi+\sigma$ plasmon energies were red-shifted for uniaxial and biaxial strains and blue-shifted for compressive strains. The π-plasmons disappeared for tensile and asymmetrical strains. Biaxial asymmetrical strains had no effect upon interband transitions and $\pi+\sigma$ plasmon energies.

Thermal transport in silicene was studied[101] by means of the non-equilibrium molecular dynamics simulation of the effect of tensile strain and isotopic doping upon thermal conductivity. The conductivity of the silicene was only some 20% of that of bulk silicon. It increased under small tensile strains, and then decreased at large strains. Isotopic doping led to a U-shaped change in the conductivity as the isotope concentration ranged from 0 to 100%. Ordered doping, leading to an isotope superlattice, led to a much greater reduction in conductivity than did random doping.

The strain-dependent thermal conductivity of monolayer silicene under uniform biaxial tension was again calculated[102] by solving the phonon Boltzmann transport equation, using interatomic force constants deduced from first-principles calculations. The conductivity could greatly increase with strain and, depending upon the size, the conductivity of strained material could be several times higher than that of unstrained material. This strain-dependence was attributed mainly to a sharp increase in the acoustic phonon lifetime due to flattening of the structure during stretching; something which is unique to silicene among 2-dimensional materials.

Thermal transport in polycrystalline silicene was examined[103] by using a multi-scale method in which non-equilibrium molecular dynamics simulations were used to assess the interfacial thermal resistance of grain boundaries with regard to tensile strain and temperature. The room-temperature Kapitza conductance at grain boundaries was deduced to be about 2.56×10^9 or $2.46 \times 10^9 W/m^2K$ when using the Tersoff and Stillinger-Weber interatomic potentials, respectively. Regardless of the temperature, which did not affect the Kapitza resistance, the interfacial thermal resistance could be controlled by straining. By varying the grain size, the conductivity could be changed by up to an order of magnitude.

Molecular dynamics simulations showed[104] that biaxially-stretched monolayer silicene exhibited an increase in conductivity of up to 10 times as an imposed strain ranged from zero to 0.1. This was attributed to increasing contributions arising from long-wavelength phonons. Subsequent increases in strain from 0.11 to 0.18 led to a plateau in conductivity via an oscillatory route which was governed by dynamic bonding under extreme loading conditions.

The temperature- and size-dependences of the lattice thermal conductivity of monolayer silicene under tensile straining were deduced[105] from first-principles calculations of the phonon Boltzmann transport equation within the framework of 3-phonon scattering theory. This again showed that the conductivity exhibited an obvious strain-dependence. The TA phonon-mode was responsible for most of the conductivity, while that of the ZA mode was repressed. With increasing tensile strain, the contribution of the LA-mode to conductivity rapidly increased and eventually became the dominant one. This was tentatively attributed to a reduction in the warping of the silicene and to a restoration of crystal symmetry by the tensile strain. When the characteristic size was less than 10nm, the conductivity became insensitive to temperature and, with increasing tensile strain, the effective phonon mean-free-path of the silicene also increased, thus rendering the size-effect more obvious.

Non-equilibrium molecular dynamics simulations were used[106] to study the thermal conductivity of silicene nanotubes with regard to the effects of strain, vacancies, grain boundaries and temperatures of 300 to 700K. It was found that the conductivity varied, with nanotube length, from 24 to 34W/mK but did not depend upon the diameter or chirality. Nanotubes with a grain boundary had an almost 30%-lower conductivity than that of boundary-free ones. Phonon power spectral density calculations indicated that increasing the defect concentration and temperature decreased the conductivity. The effect of strain was non-linear, in that the conductivity first increased markedly with tensile strain and then decreased. The maximum conductivities of armchair and zig-zag edge tubes occurred at strains of about 3 and 5%, respectively; that is, some 28% higher than that of unstrained material.

First-principles calculations were used to study[107] strain-modulated phonon transport in 2-dimensional pentagonal silicon and hexagonal silicene structures. These were hydrogenated in order to impart thermodynamic stability. In pentagonal silicene structures, the conductivity initially exhibited a marked improvement as the tensile strain was increased from 0 to 10%, but then stabilized at strains greater than 10%. In-plane modes contributed most to the group-velocity increase during the straining of penta-SiH and the phonon group-velocity and phonon-lifetime increased with applied strain. There was a similarity of the pentagonal and hexagonal silicene responses, in spite of the differences in geometry: differing out-of-plane flexural modes resulted in differing conductivity-sensitivities to strain. The disparity in anharmonicity then led to differing thermal conductivities under zero strain.

The lattice thermal conductivity of free-standing silicene under various isotropic tensile strains and temperatures was determined[108]. There was a marked size-dependence of the conductivity under tensile straining; it diverged with increasing system-size. The intrinsic room-temperature conductivity for unstrained material converged, with system-size, to 19.34W/mK at 178nm. The room-temperature conductivity of strained silicene became as large as that of bulk silicon at 84µm. In large samples of silicene, the conductivity increased with tensile strain, with a peak at 7%, and then decreased.

A first-principles study was made[109] of the effect of biaxial strains of up to 5%, under a

Table 13. Effect of biaxial tensile strain upon silicene

Strain(%)	Temperature(K)	Elastic Constant(N/m)	Conductivity(W/mK)
0	300	41.31	0.99
4	300	27.34	0.93
8	300	17.25	0.74
14	300	4.52	0.62
0	500	-	0.67
4	500	-	0.62
8	500	-	0.51
14	500	-	0.42
0	700	-	0.49
4	700	-	0.44
8	700	-	0.36
14	700	-	0.28

perpendicular electric field, upon the properties of low-buckled silicene. With increasing biaxial strain, the conduction bands at the high-symmetry Γ and M points of the first Brillouin zone shifted appreciably towards the Fermi level. The field opened a small band-gap at the K-point. The field-induced gap-opening could be increased by a compressive strain and reduced by a tensile strain.

The effect of biaxial tensile strains of up to 14% upon the properties of SiH monolayer was studied (table 13)[110]. Distortion of the structure by the strain changed the characteristics of the electronic band structure such as to produce an indirect-to-direct band-gap transition together with a decrease in the band-gap, effective mass and relaxation-time of carriers. The Seebeck coefficient fell to a minimum value at 14% strain, while the electrical conductivity and electronic thermal conductivity attained maxima at 14% strain. The tensile strain led to an appreciable decrease in lattice thermal conductivity; at a strain of 14%, the lattice thermal conductivity at 700K decreased by some 43% as compared to its unstrained value.

Stone-Wales Defect

During the synthesis of ultra-thin materials possessing an hexagonal lattice structure, Stone-Wales defects are quite likely to be formed (figure 3). The existence of such topological defects in a graphene-like structure produces marked changes in their electronic and mechanical properties. The energy barrier for the formation of Stone-Wales defects in free-standing silicene is about 2.4eV, and the barrier for Ag(111)-supported silicene is about 2.8eV; significantly lower than the value (about 9.2eV) for graphene. The buckled nature of silicene presents a high energy barrier to the healing of the Stone-Wales defect, and defective silicene is therefore stable even at high temperatures. Silicene containing Stone-Wales defects is semiconducting, with a direct band-gap of 0.02eV. This value depends upon the concentration of the defects. Nitrogen substitution in Stone-Wales defected silicene indicates that the defect lattice sites are the least-preferred substitution locations for nitrogen atoms. The easy formation of Stone-Wales defects in silicene should provide a guideline for band-gap engineering in silicene-based materials[111].

Investigation of the structural and electronic properties of Stone-Wales defects in silicene sheet and nanoribbons, using periodic boundary conditions and first-principles calculations, leads to a good agreement with available values obtained using earlier first-principles calculations. For all of the concentrations considered, the Stone-Wales defects prefer to locate themselves at the edge, due to the lower formation energy there. The Stone-Wales defects at the centre of silicene nanoribbons slightly influence the electronic properties, whereas Stone-Wales defects at the edge of nanoribbons split the degenerate edge states and introduce a sizable gap which depends upon the concentration of the defects.[112] The Stone-Wales defects produce a repulsive perturbation potential, leading to charge on the edge states at the side with the defect transferring to the side without the defect.

First-principles density functional theory calculations of the formation energy and kinetic barrier of the Stone-Wales defect, in infinite and finite sheets of silicene, show that the formation energies and barriers in both cases are significantly lower than those for graphene. Compared with infinite sheets, the energy barriers and formation energies are lower for finite sheets. On the other hand, due to the low barriers the defects are expected to heal out of finite sheets[113].

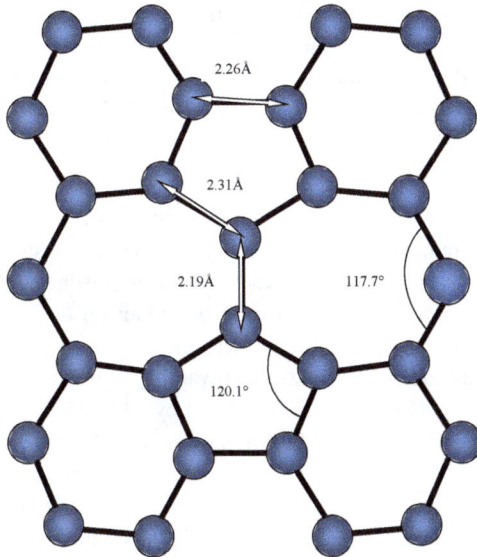

Figure 5. Stone-Wales defect in silicene. Note the changes in bond-angles and atomic spacings as compared to those of the basic silicene structure.

Density functional theory and non-equilibrium Green's function techniques show that, in zig-zag silicene nanoribbons containing Stone-Wales defects, the predicted formation energy is significantly lower than that of graphene and silicene; implying a high stability of such a defect in nanoribbons. A negative differential resistance is observed, within certain bias-voltage ranges, for both perfect and Stone-Wales defected nanoribbons[114]. Other results, obtained by applying these methods to armchair silicene nanoribbons containing a Stone-Wales defect, an interior/edge vacancy and an edge dangling bond, show that structural imperfections can alter the electronic structure (energy bands, density of states) of nanoribbons and either increase or decrease the ballistic current[115]. *Ab initio* calculations of Stone-Wales defects, single and double vacancies, and adatoms show that Stone-Wales defects can be effectively recovered by thermal annealing. Single vacancies have a much higher mobility than that of double vacancies, and two single vacancies are very likely to coalesce into a single double vacancy in order to lower the overall energy. The existence of Stone-Wales and double vacancies can introduce small gaps into silicene, while that of a single vacancy can transform semi-metallic silicene into the metallic form. Adatoms are unexpectedly stable and can markedly affect the electronic properties. In particular, silicon adatoms as self-dopants in silicene sheets can induce long-range spin polarization as well as an appreciable band-gap; yielding an all-silicon magnetic semiconductor[116].

Molecular dynamics finite-element methods, using the Tersoff potential, have been used to predict the Young's modulus, Poisson ratio and uniaxial tensile stress–strain curves for the armchair and zig-zag directions. These show that pristine and slightly defected silicene sheets exhibit almost the same elastic behaviour up to fracture. However, the presence of a single defect significantly weakens the sheet and results in a considerable reduction in the fracture strength. One two-atom vacancy at the sheet's centre reduces the fracture stress by 18 to 20% and the fracture strain by 33 to 35%. The weakening effect of Stone–Wales defects depends upon the tensile direction and upon the orientation of the defect[117]. First-principles calculations of the properties of zig-zag silicene nanoribbons containing Stone-Wales defects reveal that two types of Stone-Wales defect, SW-I and SW-II, exist in the nanoribbons. The defect is found to be most stable at the edge of the nanoribbon, regardless of orientation. It is even more stable here than it is in an infinite silicene sheet. In addition, the zig-zag silicene nanoribbons can transform from semiconductors to metals or semi-metals upon modifying the Stone-Wales defect location and concentration. For a given defect concentration, the band structures which are affected by the SW-I defect are more distinct than those which are affected by the SW-II defect, when the defect is at the edge. These results suggest the possibility of tuning the electronic properties of zig-zag silicene nanoribbons by using Stone-Wales defects[118].

Germanene

As with silicene, the amazing electronic and opto-electronic properties of free-standing graphene have led to the consideration of other possible 2-dimensional materials that might be easily integrated with the present generation of electronic technology. Unlike 2-dimensional oxides and dichalcogenides, the elemental 2-dimensional analogues of graphene, are promising candidates.[119] Atomically thin 2-dimensional materials offer the ultimate thickness-scaling of nano-electronic devices and this could substantially improve electrostatic control and permit a number of low-power applications.[120]

Preparation

Flat 2-dimensional germanium sheets having a honeycomb lattice which matches that of germanene have been prepared by depositing sub-monolayers of germanium onto graphite at room temperature and then annealing them at 350C. Scanning tunnelling microscopy shows that the germanene islands exhibit small buckling, with no atomic reconstruction, and do not seem to undergo alloy formation or hybridization with the substrate. Density functional theory calculations of the structural properties agree well with experimental results, and indicate that the germanene sheet interacts only weakly with the substrate. Band-structure calculations confirm that the Dirac cone of free-standing germanene is preserved in layers supported by graphite.[121] The germanene islands undergo slight, but characteristic, charge transfer with the graphite substrate. This is predicted by *ab initio* simulations and is again in excellent agreement with scanning tunnelling spectroscopic measurements.

Germanene has been successfully synthesized on molybdenum disulfide, a band-gap material. Pre-existing defects in the disulfide surface act as preferential nucleation sites for the germanene islands. The lattice constant of the germanene layer (3.8Å) is some 20% greater than the lattice constant (3.16Å) of the disulfide substrate.[122] Scanning tunnelling spectroscopy and density functional theory calculations reveal that, in addition to the linear dispersion bands at the K points, there are two parabolic bands which cross the Fermi level at the Γ point.

Monolayer germanene has been successfully prepared on an Sb(111) surface by means of molecular beam epitaxy. Scanning tunnelling microscopy reveals a dendritic structure at low germanium coverages, and mosaic patterns at high coverages; both having a local 1 x 1 lattice. First-principles calculations confirm the 1 x 1 low-buckle structure of germanene.[123] The dendrite and mosaic patterns are attributed to strain modulation introduced by the large lattice mismatch between germanene and the antimony substrate.

Monolayer germanene can be grown on metallic substrates but, when lying directly on a metal surface, the reconstructed atomically-thin sheets are prone to lose their massless Dirac fermion character and the unique associated physical properties of the free-standing material. Few-layer germanene, created via dry epitaxy on a gold template, possesses Dirac cones thanks to a reduced interaction. This finding establishes, using synchrotron-radiation photo-emission, scanning tunnelling microscopy imaging and surface electron diffraction, that few-layer germanene is one of the rare 2-dimensional Dirac materials. Because germanium is already being used in the mainstream silicon-based electronics industry, the possibility of using germanene to scale-down to beyond the 5nm node is very appealing. Another interesting property which might be within reach is a robust quantum spin Hall effect for application in spintronics, together with the engineering exploitation of Floquet-Majorana fermions – using light - for quantum computing.[124]

Atom-thin ordered 2-dimensional multiphase films can be grown *in situ* via germanium molecular beam epitaxy onto a gold (111) substrate; analogous to the formation of silicene layers on silver (111) templates. One phase, which forms large domains as observed using scanning tunnelling microscopy, has a clear almost-flat honeycomb structure. Using synchrotron radiation core-level spectroscopy and density functional theory calculations, this phase can be identified as being a $(\sqrt{3} \times \sqrt{3})R(30°)$ germanene layer combined with a $(\sqrt{7} \times \sqrt{7})R(19.1°)$ Au(111) super-cell. The above is therefore a reliable process for synthesis on gold.[125] A more detailed examination has been made of the electronic structure of monolayer materials based upon this double hexagonal geometry with (1 x 1) and $(\sqrt{3} \times \sqrt{3})R30°$ superstructures. Inspired by the 2-dimensional root system of an exceptional Lie algebra, this hexagonal atomic configuration is treated as involving two hexagons of unequal side-length set at an angle of 30°. The principal hexagonal unit cell contains 12 atoms instead of the usual configuration which involves only 6, relying only on the (1 x 1) superstructure. *Ab initio* numerical calculations of graphene and germanene with their double hexagonal geometry showed that the usual electronic properties and lattice parameters of such materials are modified: that is, the lattice parameters are increased. In the single hexagonal geometry, germanene behaves like a semimetal. In the double hexagonal geometry it becomes metallic.[126] When fabricated on a Pt(111) surface, germanene exhibits a buckled configuration with a (3 x 3) superlattice that coincides with the $(\sqrt{19} \times \sqrt{19})$ superstructure of the substrate. Covalent bonds exist throughout the germanene layer.[127] Stable films of germanene can thus be grown on platinum and gold substrates. A germanium adatom, adsorbed on germanene, buries the host germanium atom beneath it to form a dumb-bell structure; an exothermic process which occurs spontaneously. The attractive dumbbell-dumbbell interaction then favors a high coverage of dumb-bells. This presages the existence of stable new germanene phases which are constructed from a periodically repeating coverage of dumb-bell structures which exhibit diverse electronic and magnetic properties.[128] Calculation of the stable configuration for planar, and low (<1Å) buckling configurations, and analysis of the electronic band structure, reveals a linear band dispersion giving rise to massless Dirac fermions with a Fermi velocity that is about two-thirds that of graphene. The monolayer stability is shown to be directly attributable to the phonons present, with the instability being driven by the out-of-plane ZA and ZO phonon modes. Long momentum relaxation-lengths and high carrier mobilities are predicted to exist in germanene-based devices, as carrier relaxation via phonon scattering is inhibited because the electron-optical phonon coupling matrix elements are expected to be small; a factor of about 25 times smaller than in graphene.[129]

First-principles calculations of germanene/graphene bilayers, assuming various super-cell models in order to reduce the strain in the lattice-mismatched bilayer systems, suggest – following energetics analysis and electronic structure deliberations - that graphene can be used as a substrate for the synthesis of monolayer germanene. Multiple phases consisting of single-crystalline germanene having various orientations relative to the substrate can coexist at room temperature. The weak interaction between the overlayer and the substrate preserves the low-buckle structure of germanene, as well as its linear energy bands.[130]

Germanene layers having a Lonsdaleite structure were first synthesized from SiGe thin film by using a N_2 plasma-assisted process. Multi-layered germanene can then be directly observed, and the derived lattice parameters are almost consistent with theoretical predictions. Large-scale multi-layered germanene can be used in various applications.[131]

It remains a challenge to obtain germanene with preserved Dirac electrons. The so-called sandwich-dehydrogenation approach involves fabricating germanene by dehydrogenating germanane within a sandwich structure. Dehydrogenation can spontaneously occur in such a structure, and overcomes the problem of the amorphization which occurs during the heating dehydrogenation approach. The germanene obtained preserves the Dirac electronic properties very well. The Fermi velocity of germanene can moreover be efficiently manipulated by controlling the interlayer spacing between the germanane and the sandwich surfaces. These results offer guidelines to the fabrication of perfect 2-dimensional materials.[132]

Using first-principles calculations, variations in the band-structure due to straining were investigated[133]. Germanene exhibited a strain-induced self-doping behaviour which was linked to its buckled structure and could not, for example, arise in graphene. Under a compressive strain, the Dirac-point moved below the Fermi level and nanosheets behaved as if n-type doped. Under tensile straining, p-type doping was introduced by moving the

Dirac-point to above the Fermi level. That is, n-type/zero-bandgap and p-type semiconducting features could be switched by straining.

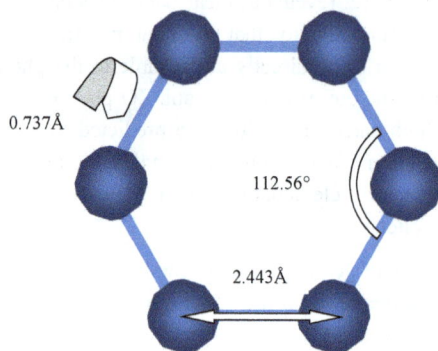

*Figure 6. Basic germanene hexagon, showing Ge-Ge
bond-length, bond-angle and buckling*

Structure

Computational techniques have been developed in order to construct 3-dimensional statistical density maps, and thus to be able to view directly the local atomic structure of amorphous germanium and aid the interpretation of various state-of-the-art experiments and simulations. The continuous random network model has been compared with an experimental model, as refined using reverse Monte Carlo methods. In the refinement, amorphous germanium admits of two dominant structures: a fourfold-coordinated tetrahedron and a buckled threefold-coordinated local structure, similar to that of silicene and germanene. Such structures account for 95.7% of the total number of atoms, in a 5:2 ratio. The present method reveals well-defined structural ordering in the second shell of amorphous germanium. This method of visualization allows the interpretation of complex disordered materials and reveals the bimodal structures of amorphous germanium.[134]

Germanene shares many similarities with graphene but, unlike the planar graphene lattice, the germanene honeycomb lattice is buckled and comprises two vertically-displaced sub-lattices. Indeed, the heavier analogues of graphene, such as silicene and germanene, all tend to be buckled. The buckling leads to interesting properties such as a direct band-gap in hydrogenated germanene. It is interesting that the sequential replacement of carbon by germanium in benzene leads to increasing buckling, with a maximum buckling distance of 0.61Å being found for Ge_6H_6. The origin of a buckling-induced lowering of the symmetry, $D_{6h} \rightarrow D_{3d}$, can be traced to pseudo Jahn-Teller distortion along the b_{2g} normal mode, arising from mixing of the non-degenerate (A_{1g}) ground state with the low-lying ($\Delta_0 = 4.36eV$) excited state of B_{2g} symmetry. Buckling also leads to an increased chemical reactivity of germanene; especially with hydrogen, to yield germanane. The high affinity of germanene for hydrogen explains a reported experimental synthesis of exfoliated layers of germanane. Formation of the latter leads to the opening-up of a large band-gap; thus making hydrogenation a chemical route for the control of electronic properties in these 2-dimensional materials. The presence of buckling in germanene also leads to greater hole reorganization energies than for polyaromatic hydrocarbons having the same nuclearity.[135]

First-principles calculations of the structural and electronic properties of germanene, and structural optimization, reveal that buckled germanene is more stable than its planar counterpart by about 0.35eV. Compared to planar graphene, with its buckling parameter of 0Å, the germanium sheet is buckled by 0.737Å but has a similar electronic structure with a zero band-gap at the K point. Further investigations of the effects of complete hydrogenation, and consideration of various geometrical configurations (chair, boat, table, stirrup), show that the chair-like structure has the highest binding energy per atom. The hydrogenation of germanene yields germanane, with a direct band-gap of 1.8eV.[136]

Density functional theory calculations reveal that free-standing germanene is a 2-dimensional Dirac fermion system. That is, the electrons behave like massless relativistic particles which can be described by the Dirac equation; the relativistic variant of the Schrödinger equation. The spin-orbit gap in germanene is about 24meV, which is much larger than that in graphene: that is, less than 0.05meV. This makes germanene an ideal candidate material for exploiting quantum spin Hall effects at experimentally accessible temperatures. The germanene lattice also offers the possibility of opening-up a band-gap by means of, for example, the imposition of an externally applied electric field, the adsorption of foreign atoms or the use of a suitable substrate. Such opening-up of the band-gap permits the preparation of germanene-based field-effect devices. It is necessary therefore to find means of synthesizing germanene and to predict its properties, using mainly density functional theory calculations.[137]

High-resolution photo-emission and density functional theory calculations of the core-level aluminium 2p and germanium 3d lines, found for germanene grown onto Al(111) by using synchrotron radiation, show that the germanium 3d line is shifted towards low binding energies - with respect to bulk germanium - and has 3 components; reflecting the site multiplicity of the germanene atomic structure. The calculations also reveal appreciable sizable charge localization at the germanene/Al(111) interface, charge transfer from the aluminium surface atoms to the germanene and the existence of 3 non-equivalent germanium sites having 3 different atomic Bader charges; in agreement with the photo-emission measurements.[138]

Structural analysis of germanium layers, deposited by molecular beam epitaxy onto Ag(111) surfaces - with and without an AlN buffer layer - and investigated using X-ray absorption spectroscopy at the Ge-K edge, shows that - for germanium layers deposited onto an h-AlN buffer layer on Ag(111) - an interatomic Ge-Ge distance of 2.38Å exists. This is typical for 2-dimensional germanium layers and is in agreement with theoretical predictions for free-standing slightly-buckled germanene. First-principles calculations, performed within the density functional theory framework, support experimental RHEED and X-ray absorption spectroscopy data, and thus provide evidence for epitaxial 2-dimensional germanium layer formation on h-AlN/Ag(111) templates.[139]

First-principles density functional theory investigation of surface functionalization and strain effects upon full- and half-passivated germanene shows that fully-passivated germanene has a different band structure. That is, the band-gap of GeH is larger than that of GeF or GeCl. When both surface passivation and strain are applied, the germanene undergoes a transformation from semiconducting to semi-metallic. Half-passivated germanene also has a different magnetic character. That is, Ge_2H is a ferromagnetic semiconductor while Ge_2F and Ge_2Cl are anti-ferromagnetic semiconductors. The stability of the magnetic coupling of Ge_2X-type materials can be modulated by applying an external strain. The calculations thus indicate that the electronic and magnetic properties of passivated germanenes depend strongly upon surface functionalization and strain.[140] First-principles calculations of a new type of hybrid phase which comprises buckled germanene, with saturated or half-saturated degrees of alkali metal atom adsorption, similarly suggest that adsorbed lithium, sodium or potassium can be used to synthesize germanene-based new phases in two dimensions. The predicted new phases, Ge_2X_2 or Ge_2X_1 (X = Li, Na, K) on a single germanene sheet could exist at room temperature. The formation energy of the Ge_2Li_2 configuration, resulting from complete lithiation, is even more favorable than that of germanane. Charge transfer is significant between the alkali metal atoms and the germanium; reflecting the ionic interactions between them, and charge-density analysis further indicates that a covalent component exists to some extent in two-dimensional Ge_2X_2 and Ge_2X_1 (X = Li, Na, K) phases. Complete lithiation of germanene even leads to a semiconductor with an energy gap of 0.14eV. The 2-dimensional phases of Ge_2X_1 (X = Li, Na, K) are metallic, with weak polarization on the Fermi level and in unoccupied states. Half-lithiated germanene has a local magnetic moment of $0.48\mu_B$ on germanium atoms with neighbouring lithium adatoms.[141]

Density functional theory studies of the stability of germanene under biaxial tensile strain, and of associated modifications of the electronic properties and phonon spectrum, show that the germanene lattice is stable at up to 16% strain; where the Dirac cone shifts towards higher energies and hole-doped Dirac states are created. The latter is due to weakening of the Ge-Ge bonds and to a reduction in the s-p hybridization. The calculated Grüneisen parameter exhibits a similar dependence upon strain.[142]

An hierarchical first-principles investigation of the entangled effects of lattice dimensionality and bond characteristics on the lattice dynamics of germanene shows that bond-bending contributes negatively to the Grüneisen constant, γ, resulting in a negative acoustic γ. Stretching however contributes positively to γ, resulting in a positive optical γ. Layer-thickening caused by chemical functionalization tends to increase the acoustic γ, due to the increased bond-stretching effect. On the other hand, bond-weakening caused by chemical functionalization tends to decrease the optical γ, due to the decreased bond-stretching effect. Excitation of the negative-γ modes results in a negative thermal expansion, while mode excitation and thermal expansion compete with each other in governing the material's thermomechanics. The sensitive structural and electronic responses to functionalization help to deduce a generic physical picture of 2-dimensional lattice dynamics.[143]

Topological phases appear in silicene and germanene due to strong spin-orbit interaction in an external perpendicular magnetic field. Below and above a critical field of 10T, silicene under 3% tensile strain[144] exhibited quantum spin Hall and quantum anomalous Hall phases, respectively. A little above the critical field, an energy-gap in the meV-range was obtained indicating that a quantum anomalous Hall phase could be created in silicene. This is not possible in graphene or germanene.

Scanning tunneling microscopy, low-energy electron diffraction, and angle-resolved photoemission spectroscopy have suggested[145] that two distinct phases associated with germanene can be grown on Ag(111). One has an alternately buckled full honeycomb lattice which is tensile-strained and partially commensurate with the substrate so as to form a striped phase. The other is a quasi-freestanding phase which is also consistent with a honeycomb lattice having a lattice-constant which is incommensurate with the substrate but which is very close to the theoretical value for free-standing germanene. The striped phase, with its lower atomic density, can be driven into the other phase and coexist with the latter upon additional germanium deposition. Band-mapping and first-principles calculations reveal that an interface state exists only in the striped phase, but the characteristic σ-band of free-standing germanene appears only in the other phase. It is concluded that adlayer/substrate commensurability plays a key role in affecting the electronic structure of germanene.

The electronic and optical properties of armchair germanene nanoribbons were studied via first-principles calculations[146]. Upon increasing the ribbon width the band-gap of pristine samples decreased along three different paths according to the criteria: n = 3P, n = 3P+1 and n = 3P+2, where n was the number of germanium atoms across the width and P is an integer. All three types are direct band-gap materials and the magnitude of the band-gap is in the order: (3P+2) < (3P) < (3P+1). The electronic and optical properties of the ribbon materials can be varied over a wide range by applying a tensile strain.

Mechanical Properties

First-principles density functional theory calculations have been used to deduce the mechanical properties of single-layer and free-standing germanene. Uniaxial tensile and compressive simulations were carried out to probe and compare stress-strain properties; such as the Young's modulus, Poisson ratio and ultimate strengths (table 14).[147] Others have found a similar value (42.05GPa.nm) for the Young's modulus of the armchair form.[148]

Table 14. Mechanical properties of germanene

Sample	Property	Value
armchair	Young's modulus	44GPa.nm
zig-zag	Young's modulus	43.4GPa.nm
armchair	Poisson ratio	0.29
zig-zag	Poisson ratio	0.35
armchair	UTS	4.7GPa.nm
zig-zag	UTS	4.1GPa.nm
armchair	ε_{UTS}	0.2
zig-zag	ε_{UTS}	0.205

The stability of germanene under biaxial tensile straining was studied[149] using density functional theory. The phonon spectrum showed that, at up to 16% strain, the germanene lattice was stable, where the Dirac-cone shifted towards a higher energy and hole-doped Dirac states appeared. The latter change was due to a weakening of the Ge-Ge bonds and reduction in s-p hybridization. The calculated Grüneisen parameter exhibited a similar dependence upon strain as that reported for silicene, but this was different to that of graphene.

First-principles density functional theory calculations were used[150] to investigate the mechanical properties of single-layer and free-standing germanene. Uniaxial tensile and compressive simulations were used to compare the Young's moduli, Poisson ratios and ultimate strengths (table 15). Loading in the armchair direction led to higher elastic moduli and tensile strengths, but to a lower Poisson ratio and smaller ultimate tensile elongation, than those for loading in the zig-zag direction. It was predicted that armchair loading led to the elongation of all bonds whereas zig-zag loading affected only the length of bonds which were partially oriented along the loading direction. Loading might not alter the lengths of bonds which were perpendicular to the tension direction. For all of the structures the buckling increased under compressive loading, while it decreased under tensile loading. This tendency was very marked for extension along the zig-zag direction at higher loadings.

Table 15. Properties of germanene sheets, predicted by density functional theory

Direction	Young's Modulus(GPa)	Poisson Ratio	UTS(GPa)
zig-zag	43.4	0.35	4.1
armchair	44	0.29	4.7

Surface Interaction

Spin-polarized density functional theory calculations, of germanene with one of 10 different adsorbed 3d transition metal atoms, show that all but one of the adatoms prefer to bind to the hexagonal hollow site. The exception is zinc, which prefers to bind to the valley site. Strong covalent bonding between the transition metal adatom and the germanene layer is found in most cases. As a result of the adsorption, the germanene can exhibit various electronic and magnetic properties, depending upon the adatom type: non-magnetic metallic (for copper adsorption), non-magnetic semiconducting (for nickel or zinc adsorption), ferromagnetic metallic (for chromium or manganese adsorption), ferromagnetic semiconducting (for vanadium adsorption) and ferromagnetic half-metallic (for scandium, titanium, iron or cobalt adsorption); with 100% spin-polarization at the Fermi level. In particular, chromium adatoms impart the largest magnetic moment, while scandium, titanium, vanadium, manganese, iron and cobalt adatoms all generate almost integer magnetic moments. The effects of the on-site Coulomb interaction, as well as the magnetic interaction between transition metal adatoms, upon the stability of the half-metallic systems are such that the half-metallic states of the scandium/germanene and titanium/germanene systems are all robust. The ferromagnetic transition-metal/germanene systems are expected to have potential application in the field of two-dimensional spintronics devices. Density functional theory studies of scandium, titanium, vanadium, chromium, manganese, iron, cobalt, nickel, copper and zinc, adsorbed on germanene, show that – again based upon the adsorption energy - the transition-metal atoms always prefer to occupy the hollow site (table 5). The resultant values of the total magnetic moment range from 0.97 to 4.95μB in going from scandium to manganese. A gap of 74meV, with a strongly enhanced splitting of 67meV, is found in case of scandium adsorption while metallic states are found in the case of titanium, chromium, manganese, iron and cobalt. Non-magnetic states are again found for nickel, copper and zinc adsorption. Semiconduction is found in the non-magnetic cases, with gaps ranging from 26 to 28meV. Vanadium-adsorbed germanene exhibits the quantum anomalous Hall effect. The results demonstrate that the transition-metal atoms and nearest-neighbour germanium atoms are ferromagnetically ordered in the case of vanadium, manganese, iron, cobalt, nickel, copper and zinc, while antiferromagnetic ordering is found for scandium, titanium and chromium. The extent of coverage of all of the transition-metal also has an effect upon the electronic structure. Ferromagnetic and antiferromagnetic coupling is detected in the case of manganese.[151] Analysis of the density of states indicates that the ferromagnetic properties of the transition-metal/germanene systems result mainly from spin-splitting of the transition metal 3d states.[152] When the structural properties of alkali-metal atoms (including lithium, sodium and potassium) adsorbed on germanene at a wide range of coverages are investigated by means of first-principles calculations, all of the atoms are found to prefer to bind to the hexagonal hollow sites of germanene. Unlike the case of graphene, the interaction between the adatom and the germanene surface is quite strong, due to the buckled hexagonal structure. With increasing adatom coverage, the binding between the alkali-metal atom and the germanene sheet weakens due to the increased adatom-adatom repulsion. Due to high charge transfer from the alkali metal to germanene, the adatom-germanium bonds which are formed are of a mainly ionic nature. Following adsorption, the semimetallic germanene becomes metallic, with its Dirac point moving below the Fermi level. The germanene thus behaves as if n-type doped. A small band-gap can open up at the Dirac point, and both the band-gap and the concentration of charge carriers in the alkali-metal/germanene system can be tuned by controlling the adatom coverage.[153] Further first-principles calculations of the adsorption characteristics of 15 different metal atoms again show that, for alkali-metal adatoms (lithium, sodium, potassium), the bonding is approximately ideal-ionic and the semi-metallic germanene finally becomes metallic; with a small band-gap opening at the Dirac point. The bonding of alkaline-earth metal atoms (beryllium, magnesium, calcium) is a mixture of ionic and covalent. Beryllium or magnesium adsorption leads to semiconducting behaviour in germanene while, as for lithium, sodium and potassium adsorption, calcium-adsorbed germanene is metallic. For most transition-metal adatoms, strong covalent bonding is found between the adatom and the germanene layer; which causes much larger distortions in the germanene lattice. Due to the partially-occupied d-orbital, the transition metals also exhibit various electronic structures upon interacting with germanene, as previously noted.[154] Other first-principles calculations of the adsorption characteristics of alkali, alkali-earth, group-III and 3d transition-metal adatoms show that the adsorption of alkali or alkali-earth adatoms has a minimal effect upon the geometry of germanene. The significant charge transfer from alkali adatoms to germanene again leads to the metallization of germanene, whereas alkali-earth adatom adsorption, where the interaction is a mixture of ionic and covalence, results in semiconducting behaviour with an energy-gap of 17 to 29meV. Group-III adatoms also bind to germanene via mixed, covalent and ionic, bonding. The adsorption characteristics of the transition metals are, as noted previously, rather complicated; though all transition-metal adsorption on germanene involves strong covalent bonding with germanene. The main contributions to this strong bonding arise from hybridization between transition-metal 3d and germanium p_z orbitals. As noted previously, depending upon the transition metal, the adsorbed systems can exhibit metallic, half-metallic or semiconducting behaviour. The dipole moment and the work function vary as a function of the adsorption energy.[155] First-principles studies have also been made of the electronic functionalization of germanene monolayers by chemisorption of hydrogen, lithium, fluorine, scandium, titanium or vanadium, with 3 general adatom-monolayer configurations being considered; each having a distinct effect upon the electronic structure and yielding metallic or semiconducting dispersions, depending upon the adatom type and configuration. The induced band-gap is an indirect or direct Γ-gap, ranging from 0.2 to 2.3eV. In general, an alternating configuration is the most energetically stable. The boat-like and chair-like conformers are degenerate, with the former being associated with anisotropic effective carrier masses. The top configuration leads to a planar monolayer and usually to a gapped dispersion. The hollow configuration, combined with vanadium adatoms, retains the Dirac cone; but with strong orbital planar hybridization at the Fermi level.[156]

First-principles density functional theory calculations show that the electronic and magnetic properties of bare and single-layer germanene with adsorbed titanium adatoms, which is charged or subjected to a perpendicular electric field, can be modified so as to impart new functionalities. When subjected to a perpendicular electric field, buckled atoms have the symmetry between their planes broken; thus opening up a gap at the Dirac points. The occupation of 3d orbitals of the adsorbed titanium atom changes due to charging or to an applied electric field; thus causing significant changes in the magnetic moment. The calculated properties exhibit however a dependence upon the size of the vacuum spacing between the periodically repeating germanene layers if they are treated using a plane-wave basis set with periodic boundary conditions. This spurious dependence can be eliminated by using a local orbital basis set.[157]

The germanene derivatives which are constructed from periodic dumb-bell units play an important role in multilayers of the honeycomb structure. First-principles calculations, based upon density functional theory, reveal the dumb-bell formation mechanisms and energetics of group-IV atoms adsorbed on graphene, silicene, germanene or stanene monolayer honeycomb structures. The stabilities of the binding structures are further confirmed, apart from stanene, by *ab initio* molecular dynamics calculations for high temperatures. Stanene instead undergoes structural instability upon adsorbing atoms. Depending upon the row-number of the adatoms and substrates, three types of binding structures are found; leading to significant changes in the electronic, magnetic and optical properties of the substrates. In particular, silicon, germanium and tin adatoms - when adsorbed on silicene or germanene - form dumb-bell structures. Such dumb-bell structures occur not only on single-layer monatomic honeycomb structures but also on compounds, including SiC and SiGe. The energy barrier to the migration of a dumb-bell structure is low, due to concerted action of the atoms. This makes dumb-bells quite mobile on substrates when preparing new single or multilayer silicon and germanium phases.[158]

Organic functionalization using molecules which contain unsaturated C-C bonds, by means of radical-initiated reactions on hydrogenated surfaces, is a favorable route for the attachment of organic layers. Such organic functionalization of hydrogenated germanene, with acetylene, ethylene and styrene, when compared with respect to adsorption on hydrogenated graphene and hydrogenated silicene surfaces, shows that the results arising from organic functionalization of hydrogenated germanene and hydrogenated graphene are very different. In the case of hydrogenated germanene, the adsorption of acetylene and ethylene is energetically favorable while the adsorption of styrene – in spite of being energetically favorable - leads to a final state whose structural configuration does not favor a chain reaction. The adsorption of these molecules on hydrogenated graphene is less likely to occur in the case of acetylene and ethylene and, for styrene, it is not energetically favorable. The differences in the surface reactivity of H-graphene and H-germanene are attributed to the greater lattice constant of hydrogenated germanene and to differences in the electronegativity of carbon, germanium and hydrogen atoms.[159]

First-principles calculations, based upon density functional theory, of the electronic properties of hydrogenated germanene show that two different atomic configurations are stable, and energetically degenerate. Upon adsorbing hydrogen, an energy-gap opens up. The energy-gap can be predicted by using the Herd-Scuseria-Emzerhof hybrid functional as well as a many-body perturbation method. The germanene is a wide band-gap semiconductor, while germanane is predicted to be a direct-gap material, regardless of its atomic configuration, with an average energy-gap of about 3.2eV. This makes the material potentially interesting for use in opto-electronic applications in the blue/violet spectral range.[160]

First-principles calculations of the electronic and magnetic properties of hydrogenated germanene show that, like silicene, buckled pristine germanene is also more stable than is pristine germanene with a planar configuration. Metallic properties can be detected in the planar structures, and semi-metallic (with zero band-gap) properties can be found in the buckled structure. As the most stable structure, buckled pristine germanene is found with hydrogen atoms in several configurations. Of the 5 configurations which have been investigated, stable structures are found only for the chair-like obtuse and boat-like obtuse configurations; both being semiconductors. Only in the chair-like obtuse configuration however does the system have a non-zero magnetic moment. Overall, as compared with 100% hydrogenated silicene and graphene, 100% hydrogenated germanene has the smallest band-gap energy.[161]

The electronic properties and scanning tunnelling microscopic topographic images of X (= F, H, O) functionalized germanene have been investigated, using density functional theory, by introducing various types of vacancy cluster and chain patterns into monolayers. The formation of vacancy clusters and chain patterns is energetically most favorable in hydrogenated germanene. The F- and H-functionalized materials are direct band-gap semiconductors, with band-gaps ranging from 0.1 to 1.9eV. The O-functionalized monolayers are metallic in nature. Upon introducing the various vacancy clusters and chain patterns, the electronic and magnetic properties are significantly modified. The F- and H-functionalized materials, with hexagonal and rectangular vacancy clusters, are non-magnetic semiconductors with modified band-gap values, while pentagonal and triangular vacancy clusters introduce metallicity and magnetic characteristics.[162] Hexagonal vacancy chain patterns induce a direct-to-indirect gap transition while zig-zag vacancy chain patterns retain the original direct band-gap nature. Calculated scanning tunnelling microscopic topographic images reveal distinct characteristics for the various types of vacancy cluster and chain patterns. This can be used to identify various vacancy patterns during functionalization.

First-principles calculations of the electronic properties of chlorine- and hydroxyl-substituted half-hydrogenated or fully-hydrogenated material show that both the concentration and the distribution of the substituent greatly affect the electronic properties. The half-hydrogenated germanene evolves, from a magnetic semiconductor to a half metal, with increasing chlorine or hydroxyl concentration. The chlorine or hydroxyl decoration of fully-hydrogenated germanene does not change the intrinsic electronic structure of the direct band-gap, but can reduce the band-gap with increasing concentration. Fully-hydroxylated and fully-chlorinated germanene are semiconductors, like fully-hydrogenated germanene. These results demonstrate that chlorine and hydroxyl treatment can tune differently the electronic properties of half- and fully-hydrogenated germanene.[163]

First-principles calculations of the intriguing electronic and magnetic properties of one-side semi-hydrogenated germanene show that, based upon phonon-mode analysis, the material is dynamically stable in the ground state. Semi-hydrogenation from one side only leads to localized and unpaired 4p electrons existing in the unhydrogenated germanium sites. This then introduces ferromagnetism into the germanene sheet, with no need for doping, cutting or etching. One-side semi-hydrogenated germanene is a semiconductor with a direct energy-gap. The gap value, as deduced using the Herd-Scuseria-Emzerhof functional, is estimated to be 1.32eV. In going from pristine to one-side semi-hydrogenated, and then to fully hydrogenated, the germanene changes from metallic to magnetic semiconducting and then to non-magnetic semiconducting. Hydrogenation is a novel means of tuning the properties, and has enormous potential for future nano-electronics use.[164]

Investigation of the stability and electronic properties of nitrogen and boron impurities in single layers of germanane, using spin polarized density functional theory within the GGA-PBE and Herd-Scuseria-Emzerhof approaches for the exchange correlation term, reveals that these impurities have lower formation energies in germanane than do their counterparts in graphane. The adsorption of hydrogen atoms in the vicinity of the defects stabilizes the system. The electronic properties of germanane can be tuned when nitrogen and boron are incorporated into the germanium network. The use of nitrogen-doping and boron-doping gives rise to n-type and p-type semiconducting properties, but the adsorption of hydrogen atoms quenches the doping effects.[165]

First-principles studies of O_2 dissociation and oxygen-atom adsorption on free-standing germanene show that, compared with the extremely active silicene in an oxygen atmosphere, germanene is less active; due to a dissociation-energy barrier of about 0.57eV. A dissociated oxygen atom then follows two opposite migration pathways on the germanene surface; quite distinct to the case of silicene. The migration and desorption of oxygen atoms at room temperature is relatively difficult, due to strong Ge-O bonding;

resulting in the formation of germanium oxides. The results reveal an interplay between germanene and O_2, and suggest an enhanced stability of germanene in an oxygen atmosphere, as compared with silicene.[166]

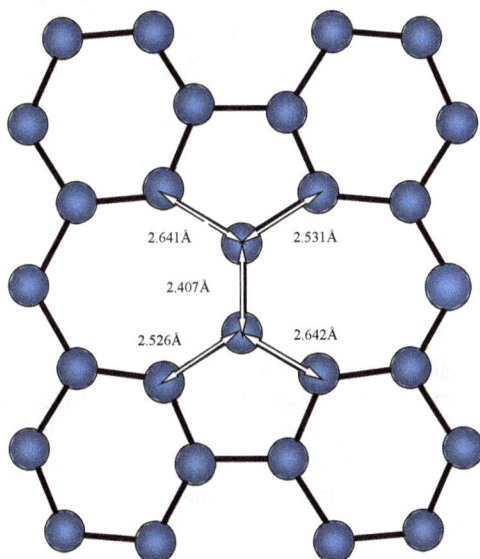

Figure 7. Stone-Wales defect in germanene. Note the changes in atomic spacings as compared to that in the germanene structure

Density functional theory studies of the adsorption of the usual gas molecules (N_2, CO, CO_2, H_2O, NH_3, NO, NO_2, O_2) on germanene show that N_2, CO, CO_2 and H_2O are physisorbed on germanene via van der Waals interactions while NH_3, NO, NO_2 and O_2 are chemisorbed on germanene via strong covalent (Ge-N or Ge-O) bonds. The chemisorption of gas molecules opens a band-gap at the Dirac point of germanene. The NO_2 chemisorption on germanene reveals strong hole-doping. Meanwhile O_2 is easily dissociated on germanene at room temperature. The differing adsorption behaviours of common gas molecules on germanene suggest a feasible means for exploiting chemically modified germanene.[167]

Materials Research Forum LLC

https://doi.org/10.21741/9781644903650

First-principles calculations of germanene with small adsorbed nitrogen-based molecules (N_2, NH_3, NO_2, NO) reveal that all of the molecules bind much more strongly to germanene than to graphene. This is due to the hybridized sp^2-sp^3 bonding of germanium atoms. The N_2, NO and NO_2 molecules all act as acceptors, while the NH_3 molecule donates electrons. Sizeable band-gaps (2 to 158meV) are opened up at the Dirac point of germanene by N_2, NH_3 or NO_2 adsorption, with only slight spoiling of its Dirac-cone shape. The NO_2 molecule also exhibits a marked p-type doping behaviour, and renders germanene metallic. With adsorbed NO molecules, germanene can change into a ferromagnetic half-metal with 100% spin-polarization at the Fermi level.[168] The various adsorption behaviours of small nitrogen-based gas molecules on germanene offer a feasible means for exploiting chemically modified germanene in a wide range of applications, such as field-effect transistors, gas sensors and spintronic devices.

Density functional theory studies of the adsorption of the ionic liquid, choline benzoate, at the surfaces of hexagonal nanosheets such as graphene and germanene show that the adsorption involves mainly van der Waals forces; leading to strong benzoate ion-surface π-stacking. There is also substantial charge-transfer from the anion to the sheet. Electronic structure analysis shows that germanium-based sheets lead to the largest changes in the HOMO and LUMO levels of choline benzoate. This provides new insights into the ability of density functional theory methods to furnish useful information concerning the adsorption of ionic liquids on nanosheets, and intimates just how ionic-liquid features might be tuned by adsorption on a suitable nanosheet material.[169]

Defects

Ab initio electronic structure and transport calculations of two-dimensional hexagonal germanium containing Stone-Wales, single vacancy (5-9-6) and two types of divacancy (5-8-5), (555-777), show that these defects constitute a local reconstruction that can be clearly identified in scanning tunnelling microscopic images. Among these defects, the Stone-Wales defect has the lowest formation energy. In the presence of structural defects, the two-dimensional hexagonal germanene maintains its Dirac-cone feature only in the case of the single vacancy. Divacancies and Stone-Wales defects destroy the linear dispersion relationship of the electrons near to the Fermi level in this system. These defects also create scattering centres, which diminish the current by some 42% in the case of the Stone-Wales or single vacancy, by 55% in the case of the 5-8-5 divacancy and by 68% in the case of the 555-777 divacancy.[170]

First-principles calculations of the structural and vibrational properties of germanene predict that a structure with a small buckling of 0.7Å and a bond length of 2.44Å is the most favored one energetically and does not exhibit an imaginary phonon mode. In calculated non-resonance Raman spectra, the highest peak is at about 290/cm. Extensive calculations of armchair germanene nanoribbons, with and without hydrogenation of the edges, reveal other Raman peaks which are distributed mainly at lower frequencies than the G-like peak that can be attributed to defects at the edges of the ribbons and are thus not present in the Raman spectra of non-defective germanene. The Raman peak which corresponds to the D mode is located at about 270/cm. The calculated G-like and D peaks are expected to be the fingerprints of the Raman spectra of low-buckle germanene.[171]

Stanene (Tinene)

Stanene is one of most important 2-dimensional materials because of its potential to exhibit room-temperature topological effects due to opening of a spin-orbit gap. Synthesis of a few layers of stanene, and atomic-scale morphological and elemental characterization confirms the presence of an hexagonal lattice of tin atoms (figure 8). The position of the Raman peak, together with the interplanar spacing, are in good agreement with those deduced from first-principles calculations, and confirm that the sheets are not merely (111) α-Sn sheets.

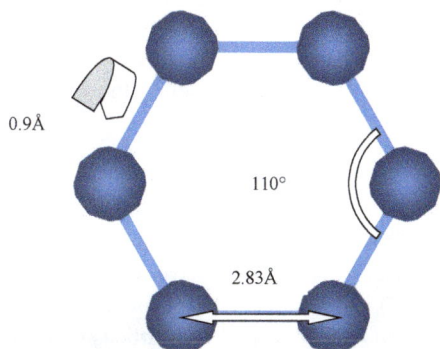

Figure 8. Basic stanene hexagon, showing Sn-Sn bond-length, bond-angle and buckling

The optical signatures at about 191nm and about 233nm, calculated using density functional theory, for low-buckle stanene are in qualitative agreement with the measured UV-Vis absorption spectrum. Atomic force microscopic measurements indicate an interlayer spacing of about 0.33nm; in good agreement with that reported for epitaxial stanene sheets. No traces of oxygen were detected, thus suggesting the absence of any oxidized phases.[172] This was also confirmed by comparative Raman measurements of oxidized stanene sheets. Stanene has also been successfully fabricated by means of molecular beam epitaxy, as confirmed by scanning tunnelling microscopy and angle-resolved photo-emission spectroscopy in conjunction with first-principles calculations.[173] Monolayer hexagonal tin, stanene, is an ideal candidate for the exploitation of graphene-type physics in nano-electronic applications. Its higher out-of-plane buckling and appreciable spin-orbit coupling lend themselves to improved Dirac-cone engineering. Investigation of the ballistic charge-transport regime of armchair tin nanoribbons with widths of 3m-1, 3m and 3m+1, where m is an integer, shows that tin ribbons exhibit earlier current onsets and carry currents which are 20% larger than those carried by C/Si/Ge-nanoribbons.[174]

Using first-principles calculations, the potential of the Ag(111) surface as a substrate for the epitaxial growth of monolayer stanene was explored. This revealed that the hexagonal structure of the stanene monolayer is well preserved on the Ag(111) surface. The height of the epitaxial monolayer is comparable to the step height of the substrate, thus permitting growth to cross the surface step and form large areas of stanene. The perfect lattice structure of free-standing stanene can be obtained when the epitaxial stanene monolayer is detached from the Ag(111) surface. The barrier to tin adatom diffusion on the Ag(111) surface is only 0.041eV, thus permitting the epitaxial growth of stanene monolayers even at low temperatures.[175]

Structure

Ab initio density functional studies of various 2-dimensional analogues show that four of them are based upon dumb-bell geometries, one on a Kagome lattice and one on the hexagonal structure. Calculations of the cohesive energies clearly show that the hexagonal structure (graphene) is the most stable one for carbon, as expected. In the case of silicon and germanium, dumb-bell structures and particularly large honeycomb dumb-bell geometries are energetically favored as compared with sp^2/sp^3-bonded hexagonal lattices such as silicene and germanene. The main reason for this is the opening-up of a band-gap in the honeycomb dumb-bell arrangements. The large honeycomb dumb-bell sheet crystals are indirect semiconductors with a K→Γ gap of about 0.5eV. In the present case, tin, the MoS_2-like symmetry is predicted to be more stable, in contrast to the stanene or large honeycomb dumb-bell geometries.[176] First-principles density functional theory predicts tinene to have a honeycomb lattice with a lattice parameter of 4.62Å and a buckling of 0.92. The electronic dispersion indicates a Dirac cone with zero gap at the

Fermi energy and a Fermi velocity of 0.97×10^6m/s. The Fermi velocity and the strength of spin-orbit interaction in stanene can be calculated within the framework of density functional theory.[177] Other considerations yield a band-gap of 0.10eV. The monolayer is thermally stable at up to 700K, according to first-principles molecular dynamics simulation, and exhibits phonon dispersion without imaginary frequencies. Control of the external electric field and the applied strain produces an induced band-gap of up to 0.21eV, whilst retaining linear dispersion but with degraded electronic transport.[178] Monolayer gray tin, tinene, is a 2-dimensional material having both Dirac characteristics and a 72meV band-gap, according to density functional calculations. As compared with silicene and germanene, tinene has a similar hexagonal honeycomb monolayer structure, but the larger buckling height of about 0.70Å. This buckling leads to phonon dispersion, without appreciable imaginary modes, indicating strong dynamic stability. It is deduced from the band structure that six Dirac cones would appear at high-symmetry K-points in the first Brillouin zone when gray tin is thinned from bulk to monolayer thickness, but a band-gap as large as 72meV could be preserved.[179]

Mechanical Properties

First-principles density functional theory calculations were used[180] to investigate the mechanical properties of single-layer and free-standing stanene. Uniaxial tensile and compressive simulations were used to compare the Young's moduli, Poisson ratios and ultimate strengths (table 16). Loading in the armchair direction led to higher elastic moduli and tensile strengths, but to a lower Poisson ratio and smaller ultimate tensile elongation, than those for loading in the zig-zag direction. It was predicted that armchair loading led to the elongation of all bonds whereas zig-zag loading affected only the length of bonds which were partially oriented along the loading direction. Loading might not alter the lengths of bonds which were perpendicular to the tension direction. For all of the structures the buckling increased under compressive loading, while it decreased under tensile loading. This tendency was very marked for extension along the zig-zag direction at higher loadings.

Table 16. Properties of stanene sheets, predicted by density functional theory

Direction	Young's Modulus(GPa)	Poisson Ratio	UTS(GPa)
zig-zag	23.5	0.42	2.2
armchair	25.2	0.36	2.6

Molecular dynamics simulations, using a modified embedded-atom method potential, show that both the fracture strength and strain of α-stanene decrease with increasing temperature. Uniaxial loading in the zig-zag direction yields a higher fracture strength and strain as compared with the armchair direction. No noticeable variation in the mechanical properties is observed for biaxial loading. At a higher loading-rate, the material exhibits a higher fracture strength and strain.[181] Theoretical study of the properties of graphene and its analogues is most complete for planar structures and for structures having small deviations from planarity. The limit of large deformations is more relevant because it governs material properties near to the point of failure. Discrete differential geometry concepts appear to be promising for linking material properties to shape, regardless of how great is the given deformation.[182] Density functional theory and simple molecular mechanics models for Young's modulus estimate the in-plane stiffness to be 40N/m.[183] Group theory analysis can be used to study the effect of strain, to understand structural phase transitions and to characterize the number of layers, crystallographic orientations and non-linear phenomena.[184]

Table 17. Mechanical properties of stanene

Material	Property	Value
armchair	Young's modulus	25.2GPa.nm
zig-zag	Young's modulus	23.5GPa.nm
armchair	Poisson ratio	0.36
zig-zag	Poisson ratio	0.42
armchair	UTS	2.6GPa.nm
zig-zag	UTS	2.2GPa.nm
armchair	ε_{UTS}	0.17
zig-zag	ε_{UTS}	0.18

First-principles density functional theory calculations have been used to investigate the mechanical properties of single-layer and free-standing material. Uniaxial tensile and compressive simulations probed and compared the Young's modulus, Poisson ratio and ultimate strength, and evaluated the effect of chirality upon the mechanical response and bond structure of the 2-dimensional substrates. The armchair and zig-zag Young's moduli are estimated to be 25.2GPa.nm (table 17).[185] Others have calculated the armchair Young's modulus to be 24.46GPa.nm.[186] Such first-principles simulations suggest that the application of uniaxial loading can change the electronic nature of the buckled structures to metallic. The elastic properties of stanene under equiaxial or uniaxial tensions along the armchair and zig-zag directions have similarly been investigated by means of first-principles calculations. The stress-strain relationship is calculated, and the relaxation of the internal atomic positions is analyzed. High-order elastic constants are calculated by fitting polynomial expressions. The Young's modulus and Poisson ratio are thus calculated to be 24.14N/m and 0.39, respectively.[187] Stanene has a lower Young's modulus than that of the preceding group-IV elements. This is attributed to the smaller sp^2-sp^3 bond energy in stanene, as compared to that in silicene and germanene. The calculated values of ultimate stresses and strains, second-order elastic constants and in-plane Young's modulus are all positive; thus proving that stanene is mechanically stable.

First-principles plane-wave calculations within spin-polarized density functional theory of the adsorption of alkali and alkaline-earth metal atoms, at various degrees of coverage, on single-layer stanene show that the most favorable adsorption sites for alkali atoms (lithium, sodium, potassium) are the hollow sites. This is similar to the behaviour of other group-IV monolayers, but the case of alkaline-earths on stanene is different to that of silicene or germanene. While magnesium and calcium are bound to stanene at hollow sites, the bridge site is energetically preferable to beryllium adatoms.[188] All of the adsorbed atoms are positively charged due to charge transfer from the adatom to the stanene monolayer. The semi-metallic bare stanene then becomes metallic, except in the case of beryllium adsorption. The latter gives rise to a non-magnetic semiconducting ground-state. These results show that stanene has a reactive and functionalizable surface similar to that of graphene or silicene. First-principles density functional theory calculations of the interaction of sodium or lithium ions on stanene films have identified the most stable binding sites and their corresponding binding energies. Bader charge analysis confirmed complete charge transfer between lithium or sodium ions and the 2-dimensional sheets.[189]

Density functional theory was used to characterize the mechanical properties of single-walled nanotubes of group-IV elements, specifically (10,10) armchair nanotubes[190]. The nanotubes were considered to be an assembly of beam elements. The Young's modulus

Materials Research Forum LLC
https://doi.org/10.21741/9781644903650

of a nanotube was calculated using finite-element methods (table 18). The Young's modulus of tin nanotubes was 83.15GPa according to density functional theory. The finite-element prediction was 82.6GPa.

Table 18. Properties of stanene nanotubes

Property	Value
Poisson ratio	0.44
flexural rigidity	2.29
E	83.15GPa
G	0.404TPa

The effect of point vacancy, bivacancy and Stone-Wales defects upon the tensile response of single-layer bulk stanene sheets was studied[191] using classical molecular dynamics simulations. The deformation of defective bulk stanene was determined by applying uniaxial tensile loading. The tensile strength and elastic modulus of stanene decreased with increasing defect concentrations. Sheets with Stone-Wales defects offered a lower in-plane stiffness. For armchair samples, the Young's modulus was 25.86N/m, the UTS was 3.45GPa and the strain at failure was 0.201.

Stanene's low tensile strength limits its use in nanodevices, but van der Waals heterostructures permit the integration of 2-dimensional layered materials. Molecular dynamics simulations have been used[192] to predict the tensile strength of Van der Waals heterostructures of stanene adsorbed on graphene, hexagonal boron nitride and silicon carbide under armchair and zig-zag loading at various strain-rates. The SiC heterostructure offered the lowest tensile strength and strain. Zig-zag loading allowed more tensile strain before fracture. The rule-of-mixtures could accurately predict the Young's modulus of heterostructures but had a limited ability to predict the tensile strength. In the case of the boron nitride heterostructure, fracture began in the stanene layer. In the graphene and SiC heterostructures, fracture began in the graphene and SiC layers under both armchair and zig-zag loading.

The tensile behaviour of planar and low-buckled stanene was analysed[193] with regard to its ideal strength in the armchair and zig-zag directions (table 19). A highly non-linear constitutive model was used in which strains of up to the fourth power were incorporated. First-principles calculations based upon density functional theory incorporated the second-, third-, fourth- and fifth-order elastic modulus tensors corresponding to planar and low-buckled states. The morphology of free-standing stanene, such as the bond-length and lattice parameter was characterised. In the case of low-buckled stanene, buckling height and dihedral angle were required. The effects of uniaxial and biaxial loading upon the latter were studied. Study of the electronic charge distribution revealed necking of the the Sn-Sn bonds under large uniaxial extension along the armchair direction.

Table 19. Ideal strength of planar and buckled stanene

Type	Direction	Strength(N/m)	Strain
planar	armchair	2.14	0.17
planar	zig-zag	1.37	0.11
buckled	armchair	3.24	0.20
buckled	zig-zag	4.05	0.21

The elastic moduli and crack-propagation behaviour of single-layer and bilayer stanene were studied[194] using molecular dynamics simulations and the Tersoff potential. Interlayer van der Waals interactions were based on the Lennard-Jones potential in the case of bilayer stanene. The effects of temperature and strain-rate upon the mechanical properties were investigated for single-layer and bilayer material in the armchair and zig-zag directions. The fracture strength and strain decreased with increasing temperature. At higher loading-rates, there was a higher fracture strength and strain. The effect of chirality upon the elastic moduli was explained in terms of a fundamental interaction between the shear modulus and Young's modulus.

Equilibrium molecular dynamics simulations were used[195] to characterize thermal and mechanical properties by using the modified embedded-atom method potential. The stress-strain behaviour of samples of various widths was analyzed under uniaxial loading. Zig-zag samples had a higher fracture strength than did armchair samples. With increasing width, the fracture strain and fracture stress of armchair samples exhibited small variations, as compared with zig-zag samples.

Defects

During the synthesis of stanene monolayers, defects are inevitably created, and always affect the properties. *Ab initio* predictions have been made of the structures, diffusion behaviour and related properties of typical point defects, such as the Stone-Wales defect, single vacancies (SV-1[55|66], SV-2[3|555]) and double vacancies (DV-1[5|8|5], DV-

2[555|777]). Scanning tunnelling microscopic images have also been simulated. The results reveal that the Stone-Wales defect can be easily recovered by annealing, due to its low reverse barrier. The SV-1[55|66] and SV-2[3|555] vacancies are the most stable. The energetically favored DV-1[5|8|5] di-vacancy can be formed by the coalescence of 2 monovacancies. The DV-2[555|777] can arise from DV-1[5|8|5] via a bond rotation which has to overcome a diffusion barrier of 0.89eV. The point defects have a non-trivial influence upon the electronic properties. The Stone-Wales defect can open up a direct gap in the energy band, without harming the high-velocity carriers.[196] Meanwhile, the SV-1[55|66] makes the stanene metallic while SV-2[3|555], DV-1[5|8|5] and DV-2[555|777] can change stanene into an indirect or direct band-gap semiconductor. Spin orbit coupling effects influence the electronic bands; especially the band-gaps.

First-principles calculations of grain boundaries in stanene reveal four typical structures having particularly low formation energies. These extended defects act as quasi 1-dimensional semiconducting or metallic wires, depending upon their geometrical structure. They are also reactive, and the adsorption of hydrogen atoms in the grain-boundary region is more stable than in bulk stanene. Adsorption of a single hydrogen atom has a drastic effect upon the electronic behaviour of the grain boundaries, and the band structure can be tuned by varying the coverage of hydrogen adsorption.[197]

Antimonene

First-principles calculations of the properties of antimonene show that the free-standing α and β allotropes are stable and semiconducting. The α-antimony form has a puckered structure, with two atomic sub-layers, while the β-antimony form has a buckled hexagonal lattice. Raman spectra and scanning tunnelling microscopic images have distinct features, which facilitates characterization of the allotropes. The β-antimony form has almost isotropic mechanical properties, while α-antimony form has strongly anisotropic characteristics. An indirect-direct band-gap transition is expected to occur when a moderate tensile strain is applied to monolayers.[198]

Preparation

The synthesis of high-quality monolayer or few-layer antimonenes has greatly hindered their development. High-quality few-layer monocrystalline antimonene polygons can be synthesized on various substrates, including flexible ones, by means of van der Waals epitaxial growth. Raman spectroscopy and transmission electron microscopy reveal that the polygons have a buckled rhombohedral atomic structure which is consistent with the theoretically predicted most stable β-phase allotrope. Very high stability of antimonenes has been observed following aging in air for 30 days. First-principles and molecular dynamics simulations confirm that, compared with phosphorene, antimonene is less likely to be oxidized and possesses a higher thermodynamic stability in an oxygen atmosphere at room temperature.[199] Antimonene polygons exhibit an electrical conductivity of up to 10^4S/m, and a good optical transparency in the visible light range.

A process which is assisted by sonification, but which does not require the addition of a surfactant, is a rapid and simple method for producing highly stable isopropanol/water (4:1) suspensions of few-layer antimonene via the liquid-phase exfoliation of antimony crystals. This method generates dispersions, of few-layer antimonene, which are suitable for on-surface isolation.[200] Atomic force microscopy, scanning transmission electron microscopy and electron energy loss spectroscopy have confirmed the formation of high-quality few-layer nanosheets having large lateral dimensions. These nanolayers are extremely stable under ambient conditions.

The pseudo-layer character of 3-dimensional bulk crystals of antimony already suggests that its 2-dimensional single-layer crystalline phase should have a buckled honeycomb structure (figure 9) like that of silicene. The antimony atoms also have an asymmetrical washboard-like structure, as does black phosphorene. *Ab initio* phonon and finite-temperature molecular dynamics calculations show that the single-layer phases are robust and can remain stable at high temperatures. Single-layer antimonene phases form bilayer and trilayer structures having wide interlayer spacings. This spacing is reduced in multilayers and the structure eventually changes into three-dimensional pseudo-layered bulk crystals. The phases are non-magnetic semiconductors, with band-gaps ranging from 0.3 to 1.5eV. Zig-zag and armchair nanoribbons of the antimonene phases have fundamental band-gaps arising from reconstructed edge states, and exhibit a range of magnetic and electronic properties, depending upon their width and edge geometry. Their band-gaps are tunable, using the width of the nanoribbon. The washboard form exhibits strongly directional mechanical properties, suggesting that strain may have a marked influence on electronic properties.[201] First-principles calculations indeed demonstrate that antimonene can be turned into a two-dimensional topological insulator by reducing the buckling height of the lattice, using tensile straining. A strain-driven band inversion in the vicinity of the Fermi level is responsible for the quantum phase transition. The buckled configuration permits the material to endure tensile strains of up to 18%, and the resultant bulk band-gap can be as large as 270meV. This tunable bulk band-gap makes antimonene a promising candidate material for achieving quantum spin Hall effects at high temperatures.[202]

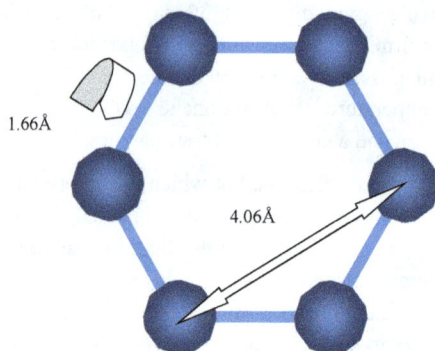

*Figure 9. Basic antimonene hexagon, showing Sb-Sb
bond-length and buckling*

Investigation of the geometrical structure and properties of antimonene doped with 3d transition metals, using spin-polarized first-principles calculations, shows that strong orbital hybridization occurs between the transition metals and antimony atoms, where covalent bonds form in the antimonene. A spin-polarized semiconducting state appears in chromium-doped antimonene, while half-metallic states appear upon doping with titanium, vanadium or manganese.[203] These results indicate that, when combined with doping states, the bands of antimonene systems exhibit a variety of features: specific dopants can lead to half-metallic characteristics, with a high spin-polarization that has potential applications in spintronics.

Mode-I stress intensity factors were estimated[204,205] for antimonene and the crack-tip displacement field which was predicted by linear elastic fracture mechanics was used to determine the stress intensity factor (table 20). Atomic-scale finite-element methods and Stillinger-Weber potentials were used to simulate tensile tests. An empirical formula was derived which related the mode-I fracture toughness to the elastic modulus, the intrinsic tensile strength, the bond-length and the buckling height of atom-thick hexagonal material.

Table 20. Properties of antimonene

Property	Value
lattice constant	4.12Å
bond length	2.89Å
buckling height	1.64Å
Van der Waals radius	2.06Å
E(armchair)	39.1N/m
E(zig-zag)	39.7N/m
Poisson ratio(armchair)	0.23
Poisson ratio(zig-zag)	0.23
maximum stress(armchair)	7.0N/m
maximum stress(zig-zag)	6.4N/m
strain(armchair)	27.6%
strain(zig-zag)	29.4%
K_I(armchair)	0.45MPa√m
K_I(zig-zag)	0.45MPa√m

Mechanical Properties

First-principles calculations were used[206] to study the ideal tensile stress-strain relationships of monoatomic monolayer 2-dimensional buckled honeycomb lattices of antimonene. The ideal strengths and critical strains were investigated for uniaxial and equi-biaxial strains (table 21). It was concluded that the critical strain was governed by the simultaneous stretching and rotation of bonds.

Table 21. Ideal strength and Young's modulus of antimonene

Property	Orientation	Value(N/m)
strength	armchair	3.75
strength	zig-zag	3.14
strength	biaxial	3.38
modulus	-	32.9

The elastic and plastic properties of 2 x 2 and 3 x 3 antimonene which was doped with scandium, titanium, vanadium, chromium, iron, cobalt, nickel, copper of zinc were determined[207]. The Young's and bulk moduli were decreased by the doping. The greatest reduction in the Young's and bulk moduli of 2 x 2 nanosheets occurred in the case of chromium- and titanium-doping. The greatest reductions were caused by vanadium and titanium in the case of 3 x 3 nanosheets. There were negligible differences in moduli between the armchair and zig-zag directions. The results for plastic behaviour showed that, apart from 2 x 2 scandium-doped samples under biaxial loading, the yield strain of all of the doped nanosheets decreased under uniaxial and biaxial loading. The greatest reduction in the yield strain of 2 x 2 nanosheets under biaxial loading occurred in copper-doped nanosheets. In the case of 3 x 3 nanosheets, the greatest reduction occurred in copper-, iron- and zinc-doped nanosheets under the same conditions. The yield strain of doped 2 x 2 nanosheets was affected by uniaxial loading, such that copper- and zinc-doped nanosheets exhibited the greatest change. In the case of 3 x 3 nanosheets, the greatest reduction occurred in chromium-doped nanosheets under the same conditions.

The effect of an external electric field upon the mechanical properties of antimonene was investigated[208] using density functional theory for uniaxial and biaxial loading. The in-plane Young's modulus and bulk modulus were calculated, and variations in an individual bond under those loading conditions were determined. The in-plane Young's modulus and bulk modulus were not greatly affected by an external electric field. The yield strain when the nanosheet entered the plastic region markedly decreased. The largest possible strain of individual bonds was not affected by an external electric field.

Nanosheets were simulated[209] by using finite-element modeling. The elastic properties were obtained by using density functional theory and properties which represented Sb-Sb bonds. The model was used to calculate the Young's modulus and critical compressive force. The simulation could closely predict the Young's modulus of monolayer antinomene. Increasing the horizontal-side length increased the stiffness of nanosheet while increasing the vertical-side length had the opposite effect. The critical compressive force of nanosheets was increased by increasing the vertical-side length while increasing the horizontal-side length had the opposite effect.

The mechanical properties of antimonene/bismuthene van der Waals heterostructures were calculated[210] by using first-principles density functional theory. Various stackings of antimonene and bismuthene were essayed in order to identify the most and least stable stackings. The Young's modulus was calculated to be 64.3N/m.

The mechanical properties of monolayer antimonene in its most stable β-phase were studied[211] using first-principles methods. The Young's and shear moduli were some 25% higher than those of bulk antimony, while the hexagonal lattice-constant of the monolayer was reduced by some 5% with respect to that of the bulk. Tensile testing under uniaxial conditions revealed strengths of 6 and 8GPa, corresponding to critical strains of 15 and 17%, in the zig-zag and armchair directions, respectively. During deformation, the structural integrity of the material was slightly better preserved in the armchair direction. The band-gap was insensitive to straining in the zig-zag direction, but straining in the armchair direction led to an indirect-to-direct band-gap transition at a strain of 4%, due to band-switching.

The effects of hydrogen, fluorine, chlorine and bromine upon the elastic and plastic properties of antimonene nanosheet were investigated[212] by using density functional theory. The Young's modulus of antimonene nanosheet under uniaxial strain and the bulk modulus under biaxial were markedly decreased by adsorption. Under plastic deformation, the yield strain of nanosheet under uniaxial straining was decreased by adsorption. On the other hand, adsorption increased the plastic strain under biaxial loading.

Classical molecular dynamics simulations and the Stillinger-Weber potential were used[213] to explore the mechanical characteristics of puckered (α-) and buckled (β-) nanotubes. All of the specimens underwent brittle failure, with a complete loss of load-bearing ability above the ultimate stress level. The brittle behaviour was unchanged by temperature and loading-rate. Due to their distinct structures in the armchair and zig-zag directions, the α-type nanotubes exhibited a distinctive anisotropic behaviour as compared with that of β-type nanotubes. The variation in elastic modulus with temperature was negligible but the tensile strength and fracture strain fell markedly at high temperatures. The effects of diameter and temperature on zig-zag material were greater, due to its lower stability.

The mechanical properties and fracture mechanisms of monolayer puckered and buckled material were determined[214] by using molecular dynamics simulations. These revealed that a stronger chiral effect led to a more anisotropic elastic behaviour in α-antimonene than in β-antimonene. The fracture toughness was deduced from the crack-line stress. The tensile strength markedly decreased with increasing temperature, crack-length and defect concentration. The elastic modulus was less affected by temperature, but was greatly affected by an increase in defect content. The strain-rate revealed a power-law relationship between strength and fracture strain.

Indiene

Early first-principles calculations of indium monolayers having a planar, puckered or buckled form showed that the planar and buckled allotropes were stable and exhibited metallic and semiconducting behaviours, respectively. Their stability and electronic properties could not be easily correlated with those of similar elemental monolayers. A van Hove singularity was observed in the electronic density of states and this could lead to an increase in the electronic conductivity. The planar allotrope remains metallic under both compressive and tensile strains, while the buckled allotrope changes from an indirect semiconductor to a metal.[215]

Mechanical Properties

Mode-I stress intensity factors were estimated[216,217] for indiene and the crack-tip displacement field which was predicted by linear elastic fracture mechanics was used to determine the stress intensity factor (table 22). Atomic-scale finite-element methods and Stillinger-Weber potentials were used to simulate tensile tests. An empirical formula was derived which related the mode-I fracture toughness to the elastic modulus, the intrinsic tensile strength, the bond-length and the buckling height of atom-thick hexagonal material.

Table 22. Properties of indiene

Property	Value
lattice constant	4.240Å
bond length	2.89Å
buckling height	1.536Å
Van der Waals radius	1.93Å
E(armchair)	8.4N/m

E(zig-zag)	8.4N/m
Poisson ratio(armchair)	0.17
Poisson ratio(zig-zag)	0.17
maximum stress(armchair)	0.77N/m
maximum stress(zig-zag)	0.72N/m
strain(armchair)	16.3%
strain(zig-zag)	17.8%
K_I(armchair)	0.065MPa√m
K_I(zig-zag)	0.063MPa√m

Arsenene

First-principles calculations indicate that two types of honeycomb structure, buckled and puckered, are stable; both possessing indirect gaps. The band-gaps of the puckered and buckled forms can be tuned by mechanical straining. Gap-closing occurs at 6% strain for the puckered form, where the bond angles between the nearest neighbours become almost equal. An indirect-to-direct gap transition occurs during straining, and 1% strain is sufficient to transform the puckered arsenene into a direct-gap semiconductor.[218] The emergence of a Dirac-like cone occurs with in-plane straining. Strain modulates not only the band-gap, but also the effective mass. The effective mass of electrons has a highly anisotropic dispersion in intrinsic puckered arsenene.[219] A bulk form of arsenic, termed gray arsenic, exists and can be used as a precursor for the buckled arsenene. Trions and bi-excitons in anisotropic 2-dimensional materials can be investigated using effective mass theory. Explicit results have been obtained for arsenene, and for materials that share its features, such as a direct quasi-particle gap and anisotropic conduction and valence bands. Trions are predicted to have markedly high binding energies and an elongated electron-hole structure, with a preference for alignment along the armchair direction where the effective masses are lower. Bi-exciton binding energies are also high, being twice those of typical monolayer transition-metal dichalcogenides.[220]

First-principles density functional theory investigations of monolayer and bilayer arsenenes under in-plane biaxial straining show that, at sufficiently high tensile strains, the monolayer changes from the buckled honeycomb form to the planar honeycomb form. The interlayer distance of bilayers nevertheless remains at about 1.371Å. Both the monolayer and bilayer forms have indirect band-gaps and the electronic properties can be tuned, as already noted, by in-plane biaxial straining. Under compressive strains, the band-gap of both monolayer and bilayer arsenene initially increases but then rapidly decreases. The monolayer form exhibits an indirect-to-direct band gap transition when the compressive strain reaches -10%. Under tensile straining, the band-gap of monolayer arsenene decreases monotonically with strain, while the band-gap of the bilayer form decreases rapidly to zero under small tensile strains.[221]

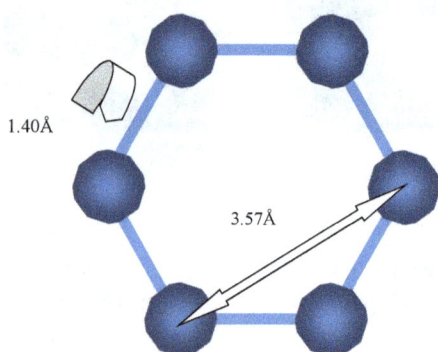

1.40Å

3.57Å

Figure 10. Basic arsenene hexagon, showing As-As bond-length and buckling

Mechanical Properties

Theoretical calculations of phonon spectra and elastic constants have verified the dynamic and mechanical stability of arsenene (figure 10). On the basis of the stable structure, the bulk modulus, shear modulus, Young's modulus and Poisson ratio have been calculated (table 9).[222] The shear anisotropy, Vickers hardness, Debye temperature, melting point and thermal conductivity have also been predicted, together with the electron mobility and relaxation time.

Mode-I stress intensity factors were estimated[223,224] for arsenene and the crack-tip displacement field which was predicted by linear elastic fracture mechanics was used to determine the stress intensity factor (table 23). Atomic-scale finite-element methods and Stillinger-Weber potentials were used to simulate tensile tests. An empirical formula was derived which related the mode-I fracture toughness to the elastic modulus, the intrinsic tensile strength, the bond-length and the buckling height of atom-thick hexagonal material.

Table 23. Properties of arsenine

Property	Value
lattice constant	3.61Å
bond length	2.51Å
buckling height	1.40Å
Van der Waals radius	1.85Å
E(armchair)	50.3N/m
E(zig-zag)	50.6N/m
Poisson ratio(armchair)	0.20
Poisson ratio(zig-zag)	0.20
maximum stress(armchair)	7.5N/m
maximum stress(zig-zag)	7.0N/m
strain(armchair)	23.1%
strain(zig-zag)	25.8%
K_I(armchair)	0.56MPa√m
K_I(zig-zag)	0.53MPa√m

Density functional theory was used[225] to study the properties of arsenene which was doped (25 or 50%) with iron, titanium or vanadium. The Young's and bulk moduli (tables 24 and 25) were determined for uniaxial and biaxial straining. The elastic modulus of doped material was smaller than that of pristine material, and the greatest and least reductions in elastic modulus of nanosheet were found after doping with titanium and vanadium, respectively.

Table 24. Young's modulus of doped arsenene

Dopant	Amount(%)	$E_{longitudinal}$(N/m)	$E_{transverse}$(N/m)
-	-	54.32	54.31
Fe	50	27.75	27.75
Fe	25	41.45	17.75
V	50	30.66	30.67
V	25	42.92	39.83
Ti	50	22.82	22.60
Ti	25	39.54	30.80

Table 25. Bulk modulus of doped arsenene

Dopant	Amount(%)	B(N/m)
-	-	25.78
Fe	50	22.66
Fe	25	18.02
V	50	19.17
V	25	10.78
Ti	50	11.69
Ti	25	22.85

The fracture properties and Young's modulus of buckled (b-arsenene) and puckered (p-arsenene) samples were investigated[226], revealing a maximum fracture-strain reduction from 0.24 to 0.14 in armchair b-arsenene when the temperature was increased from 100 to 500K. The most significant factor affecting the mechanical properties was the presence of surface defects. A maximum fracture strength reduction of 85.7% was predicted for armchair b-arsenene as the defect ratio increased from 0 to 5%. The strain rate had a negligible effect upon the mechanical properties.

The moduli of monolayer material were calculated by using finite-element methods[227]. Size had a marked effect upon the mechanical properties in that, the smaller the size, the larger were the Young's modulus and Poisson ratio. A size-change had a notable effect upon the Young's modulus in the handrail direction, but the Young's modulus of the zig-zag direction was insensitive to the size-change. Size changes cound have a marked effect upon the shear modulus in the handrail direction, whereas the shear modulus in the zig-zag direction exhibited no significant change. The effect of a size-change upon the Poisson ratio in the zig-zag direction was greater than that for the handrail direction.

The adsorption of hydrogen and fluorine atoms on arsenene nanosheet was simulated by means of density functional theory[228]. The Young's modulus under uniaxial tension was greatly degraded by adsorption (tables 26 and 27). The bulk modulus under biaxial strain was markedly reduced by adsorption. Loading was extended into the plastic regime showing that, under uniaxial tension, the plastic strain also decreased. On the other hand, hydrogen and fluorine adsorption led to an increased plastic strain under biaxial loading.

Table 26. Young's modulus of hydrogenated and fluorine-adsorbed arsenene

Hydrogenation	Absorbant	$E_{longitudinal}$(N/m)	$E_{transverse}$(N/m)
none	-	50.47	50.47
none	F	14.65	14.62
full	-	17.56	17.56
semi	-	10.61	10.61

Table 27. Bulk modulus of hydrogenated and fluorine-adsorbed arsenene

Hydrogenation	Absorbant	B(N/m)
none	-	31.11
none	F	19.59
full	-	24.84
semi	-	11.88

The mechanical properties of zig-zag and armchair buckled nanotubes were investigated[229] using density functional theory. All of the nanotubes were stable according to the vibrational spectra at room temperature. With increasing tube-diameter, the armchair nanotubes were initially found to be energetically more favourable than zig-zag nanotubes, and the strain energies of zig-zag and armchair nanotubes were similar. The Young's modulus generally increased and zig-zag nanotubes were stiffer than armchair nanotubes of comparable diameter (figure 11). The Poisson ratio was relatively insensitive to tube diameter at first, and then decreased by about an order of magnitude for zig-zag nanotubes, while the Poisson ratio of armchair nanotubes exhibited the reverse behaviour. Zig-zag nanotubes, and some armchair nanotubes of small diameter, had an indirect band-gap. Armchair nanotubes of larger diameter had a direct band-gap. The transition from indirect to direct band-gap or from semiconductor to metal could be tuned by uniaxial strain. Faceted nanotubes could be created by introducing defect lines or by conjoining various structural phases.

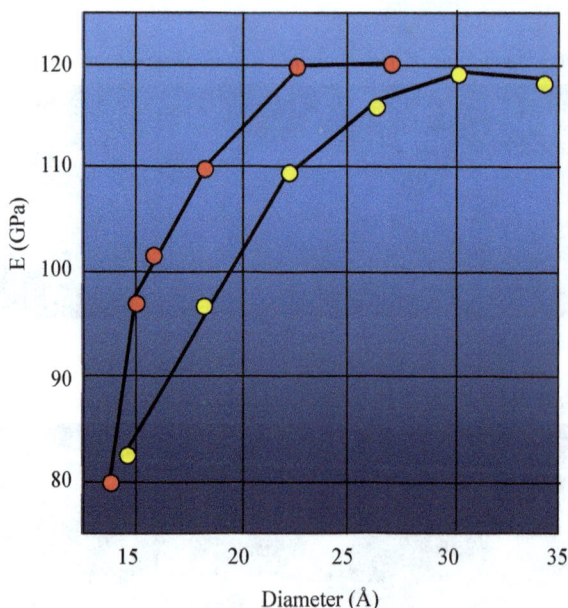

Figure 11. Young's modulus of zig-zag (red) and armchair (yellow) arsenene nanotubes as a function of tube diameter

Lithium diffusivity and mechanical properties were determined[230] for arsenene (table 28) and arsenene/graphene heterostructures (table 29) by means of first-principles calculations. Arsenene with vacancy defects offered a low diffusivity, together with low planar stiffness and poor rigidity. In order to improve the strength while retaining a low diffusion barrier, graphene was used to cover the upper arsenene layer. Arsenene/graphene heterostructures with vacancy defects offered a higher mechanical strength due to van der Waals interactions of the interlayer and to the larger Young's modulus of graphene.

Table 28. Young's modulus of defective arsenene

Vacancies	Type	E(GPa)
0	zig-zag	107.4
0	armchair	107.9
1	zig-zag	92.6
1	armchair	93.2
2	zig-zag	99.5
2	armchair	94.5

Table 29. Young's moduli of arsenene/graphene heterostructures

Composition	Type	E(GPa)
C32As71	zig-zag	833.1
C32As71	armchair	831.5
C31As71	zig-zag	782.5
C31As71	armchair	800.0
C30As71	zig-zag	709.0
C30As71	armchair	672.1
C29As71	zig-zag	707.7
C29As71	armchair	615.4
C28As71	zig-zag	628.1
C28As71	armchair	607.9

Density functional theory was used to study the effect of the adsorption of aluminium, gallium, lithium and selenium upon the elastic and plastic properties of arsenene[231]. The Young's modulus exhibited anisotropy. Apart from lithium-adsorbed material, the longitudinal Young's modulus of adsorbed materials increased while the transverse Young's modulus decreased (table 30). Apart from aluminium-adsorbed material, the bulk modulus was reduced by adsorption (table 31). The yield strain of selenium-adsorbed material under longitudinal uniaxial loading, and that of gallium- and lithium-adsorbed material under transverse uniaxial loading increased while other materials exhibited the opposite behaviour. The adsorption of gallium and lithium led to an increase in the yield strain under biaxial loading.

Table 30. Young's modulus of adsorbed arsenene

Absorbant	$E_{longitudinal}(N/m)$	$E_{transverse}(N/m)$
-	19.61	69.55
Al	35.63	62.84
Ga	31.24	47.28
Li	9.56	64.00
Se	32.85	39.73

Table 31. Bulk modulus of adsorbed arsenene

Absorbant	$B(N/m)$
-	32.76
Al	35.50
Ga	25.78
Li	27.77
Se	27.86

Density functional theory was used[232] to investigate the properties of 2 x 2 and 3 x 3 pristine material and those of cobalt-, chromium-, iron-, manganese- or nickel-doped arsenene (tables 32 to 34). Doping with the transition-metal atoms reduced the moduli of pristine samples. Isotropic behaviour was found upon comparing the Young's moduli of pristine and doped samples in the armchair and zig-zag directions. Upon increasing the applied loading, the yield strain of 2 x 2 and 3 x 3 nanosheets decreased under uniaxial loading. Under biaxial loading, the yield strain of 2 x 2 cobalt-, nickel- and titanium-doped samples increased while other 2 x 2 and 3 x 3 nanosheets exhibited the opposite behaviour.

Table 32. Young's moduli of 2 x 2 doped arsenene

Dopant	$E_{longitudinal}$(N/m)	$E_{transverse}$(N/m)
-	54.66	54.74
Co	46.43	46.38
Cr	39.06	39.07
Fe	38.55	38.67
Mn	35.13	34.71
Ni	41.27	40.91
Sc	32.38	32.48
Ti	30.43	30.28
V	36.23	36.49

Table 33. Bulk moduli of 3 x 3 doped arsenene

Dopant	$E_{longitudinal}$(N/m)	$E_{transverse}$(N/m)
-	54.69	54.74
Co	51.69	51.76
Cr	44.37	44.40
Fe	49.80	49.70
Mn	47.47	47.59
Ni	47.90	48.02
Sc	42.06	42.26
Ti	34.25	34.83
V	31.47	26.72

Table 34. Bulk moduli of 2 x 2 and 3 x 3 doped arsenene

Dopant	Type	Bulk Modulus(N/m)
-	2 x 2	30.34
Co	2 x 2	27.34
Cr	2 x 2	21.77
Fe	2 x 2	22.14
Mn	2 x 2	20.65
Ni	2 x 2	25.95
Sc	2 x 2	19.35
Ti	2 x 2	17.53
V	2 x 2	23.54
-	3 x 3	29.86
Co	3 x 3	29.08
Cr	3 x 3	26.26
Fe	3 x 3	28.74
Mn	3 x 3	24.36
Ni	3 x 3	28.63
Sc	3 x 3	24.42
Ti	3 x 3	21.69
V	3 x 3	23.92

Finite-element methods were used[233] to model arsenene nanosheets. The properties which represented As-As bonds were deduced by using first-principles calculations. The model was then used to predict the Young's modulus and fundamental frequency of nanosheets having various geometries. The model could predict the surface Young's modulus. Larger arsenene nanosheets had a greater surface Young's modulus. The vibrational characteristics were such that increasing the horizontal-side length for a constant vertical-side length led to a reduction in the fundamental natural frequency.

First-principles calculations were used[234] to study the ideal tensile stress-strain relationships of monoatomic monolayer 2-dimensional buckled honeycomb lattices of β-arsenene. The ideal strengths and critical strains were investigated for uniaxial and equi-biaxial strains (table 35). It was concluded that the critical strain was governed by the simultaneous stretching and rotation of bonds.

Table 35. Ideal strength and Young's modulus of arsenene

Property	Orientation	Value(N/m)
strength	armchair	5.15
strength	zig-zag	4.69
strength	biaxial	5.02
modulus	-	52.8

Phosphorene

So-called black phosphorus has been known, since 1960, to offer good carrier mobility as well as a fundamental band-gap. Phosphorene is the mono- or few-layer form of black phosphorus: a 2-dimensional layered material.[235] A single atomic layer of black phosphorus was experimentally prepared in 2014. It has anisotropic properties which arise from a puckered structure which is quite unlike that of graphene. Group-theory analysis of such 2-dimensional elemental systems with regard to the dependence upon the number of layers and stacking arrangements shows that, starting with the most symmetrical D_{6h1} space-group, the structures have a group-subgroup relationship. Analysis of the irreducible representations of their lattice vibrations made it possible to distinguish the various allotropes, to study the effect of strain, to understand structural phase transitions and to characterize the number of layers, crystallographic orientation and any non-linear behaviour.[236]

Preparation

Single crystals of orthorhombic black phosphorus can be grown by transport reaction, from red phosphorus, using Sn/SnI_4 as a mineralization additive. High-quality pure crystals which are several millimetres in size can be obtained, making the large-area preparation of single- or multi-layer phosphorene possible. *In situ* neutron diffraction studies show that black phosphorus is formed directly via the gas phase, without the appearance of an intermediate phase. Crystal growth is initiated upon cooling the starting materials from 500C.[237] Monolayer phosphorene can be prepared via mechanical cleavage and subsequent Ar^+ plasma-thinning. The Raman frequency of the A_{2g} mode stiffens, and the intensity-ratio of the A_{2g} to A_{1g} modes exhibits a discrete monotonic increase as the phosphorene thickness is decreased down to a monolayer. The stability of phosphorene is degraded by water, but the presence of a small amount of water permits the synthesis of high-quality material via the liquid exfoliation of black phosphorus, using dimethyl sulfoxide as a solvent. A phosphorus/water molar ratio of between 1.5 and 14 maximizes the quality of the phosphorene flakes and their stability.[238] The use of black phosphorus depends largely upon the mechanical exfoliation method, which is able to produce few-layer forms from the bulk. The exfoliation mechanism can be simulated in terms of a sliding process of bilayer phosphorene. Interlayer Coulomb interactions govern the optimum sliding pathway, revealing minimum-energy barriers as low as about 60meV and a surface energy of about $59mJ/m^2$. Considerable band-gap modulation also occurs along these sliding pathways.[239] Suitable liquid exfoliation conditions permit exfoliation to be carried out at the 10g-scale. The monolayer and few-layer flakes which are produced are crystalline and non-oxidized, and air-exposure leads to rapid oxidation and to the production of acid. When layered black phosphorus crystals are prepared by chemical vapor transport, they are oxidizable.[240] A very controllable method for fabricating high-quality air-stable phosphorene films having one or several layers is to use oxygen-plasma dry-etching to produce thick-exfoliated phosphorene flakes, layer-by-layer, with atomic precision. In such a stabilized phosphorene monolayer, it is possible to engineer defects precisely.

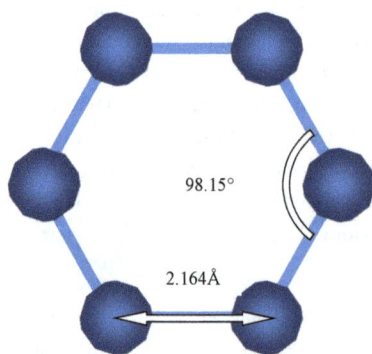

Figure 12. Basic phosphorene hexagon, showing P-P bond-length, bond-angle and buckling

Density functional theory calculations of the Raman spectra of few-layer phosphorene show that, due to the interlayer van der Waals interactions, the low-frequency rigid layer A_g breathing mode and the B_{1g} shear mode can shift by up to 45.1 and 38.5/cm, respectively, as the number of layers increases from 2 to 5. The relevant mode changes from coupled in-plane and out-of-plane vibrations in a single layer, to pure in-plane vibrations in a few-layer sample.

By using first-principles approaches, a novel kinetic pathway could be identified for the preparation of blue phosphorene via epitaxial growth. Black and blue phosphorene monolayers can be easily stabilized on Cu(111), Au(111) and GaN(001) substrates, while GaN(001) is best for fabricating blue phosphorene via half-layer by half-layer growth. The GaN(001) surface is first covered with a half-layer of phosphorus adatoms, followed by the other half.[241] Once formed, such a blue phosphorene monolayer is thermodynamically stable.

Structure

The non-planar structures of the various layered phosphorene allotropes can be mapped onto a 2-color 2-dimensional triangular tiling pattern. In the buckled structure of a phosphorene monolayer (figure 12), atoms in the so-called top positions are mapped onto dark tiles while atoms in so-called bottom positions are mapped onto light tiles. Optimum sp^3 bonding is seen throughout the structure when each triangular tile is surrounded by the same number, N, of like-colored tiles; where $0 \leq N \leq 2$. *Ab initio* density functional calculations indicate that both the relative stability and the electronic properties depend mainly upon N. This mapping approach can also be applied to phosphorene structures having non-hexagonal rings, and to 2-dimensional quasicrystals having no translational symmetry. These are predicted to be almost as stable as the hexagonal network.[242] In all of the phosphorus monolayers, the phosphorus is 3-fold coordinated, but its flexible chemistry permits it to have a coordination number of up to 6. Three new phosphorus monolayers (α-P_6, β-P_6, 558-P_6), having 2-, 3- and 4-fold coordination, are observed during epitaxial growth when flakes merge to form ridges at the boundaries. These 3 new allotropes are thermally and dynamically stable, and have an energetic stability which is comparable to that of monolayers such as δ-P, γ-P and ϵ-P. Because of their special atomic configurations, they exhibit extremely high electron mobility, anisotropic Young's moduli and optical absorbance. On the basis of *ab initio* density functional calculations, the γ-P and δ-P were suggested to be further stable structural phases of layered phosphorus in addition to layered α-P (black) and β-P (blue) phosphorus allotropes. Monolayers of some of the allotropes have a wide band-gap while others, including γ-P, undergo a metal-insulator transition which is caused by in-layer strain or a change in the number of layers. First-principles calculations also predict the existence of 7 novel phases by including 6 types of enantiomer corresponding to three types of structure which exhibit chirality. The Poisson ratio exhibited a conventional behaviour, unlike the negative values of monolayer or bulk black phosphorus, due to the structures being non-puckered. A new class of monolayer phosphorus allotropes, ε-P, ζ-P, η-P and η-P, was predicted. They were completely different to monolayer α-P (black), β-P, γ-P and δ-P, and had a buckled honeycomb lattice. The new allotropes were composed of P_4 (square) or P_5 (pentagon) units that favoured tri-coordination of phosphorus atoms. The η-P was shown to be as stable as α-P (black) and to be more stable than all previously reported phosphorene polymorphs.[243]

A 2-dimensional allotrope of phosphorus, dubbed red phosphorene, has also been reported. It results from a re-structuring of segments of blue and black phosphorene. It is dynamically stable and exhibits a thermodynamic stability which is comparable to that of black phosphorene. Because of sp^3-hybridization and the formation of a localized lone pair, red phosphorene is a semiconductor with an indirect band-gap of about 1.96eV. As usual, this can be modulated by in-plane straining, due to its wave-like configuration. Red, black and blue phosphorenes have distinct layer thicknesses, surface work functions and colors.[244]

Another allotrope, ψ-phosphorene, has been proposed which has a porous structure that is thermally and dynamically stable. Due to its unique atomic configuration, ψ-phosphorene exhibits highly orientation-dependent mechanical properties, and good flexibility. It was predicted to be semiconducting, with an indirect band-gap of 1.57eV. The electron mobility along the x-direction is up to 1.3 x 10^4cm^2/Vs; comparable to that of black phosphorene. The Au(110) surface was expected to be a suitable substrate for the synthesis of ψ-phosphorene.

Mechanical Properties

When compared with graphene and other 2-dimensional materials, phosphorene possesses a relatively low Young's modulus and fracture strength. This limits its possible applications, due to possible structural failure. Some studies of its mechanical properties have been based upon first-principles calculations at 0K. Molecular dynamics simulations for finite temperatures show that temperature has a significant effect upon the mechanical properties. The fracture strength and strain are reduced by more than 65% when the temperature is increased from 0 to 450K. The fracture strength and strain in the zig-zag direction are more sensitive to temperature than those in the armchair direction. Failure cracks prefer to propagate along the groove in the puckered structure when uniaxial tension is applied in the armchair direction. When uniaxial tension is applied in the zig-zag direction however, multiple cracks with rough fracture surfaces propagate.[245]

Molecular dynamics simulations showed[246] that temperature had a significant effect upon the mechanical properties in that the fracture strength and strain decreased by more than 65% as the temperature was increased from 0K to 450K. The fracture strength and strain in the zig-zag direction were more sensitive to the temperature change than were those in the armchair direction. The failure crack propagated preferentially along the groove in the puckered structure when uniaxial tension was applied in the armchair direction. When uniaxial tension was applied in the zig-zag direction, multiple cracks with rough fracture surfaces appeared.

Density functional theory investigations of phosphorene nanoribbons show that their ideal strength is lower than that of 2-dimensional phosphorene. The Young's modulus of armchair nanoribbons exhibits a marked size-effect, due to edge relaxations. The stress-strain curves indicate that chemisorbed oxygen atoms greatly reduce the ideal strength and Young's modulus of 2-dimensional phosphorene oxide along the zig-zag direction; especially at high oxidation ratios. Strain and oxidation have a marked effect upon phonon dispersion.[247] Molecular dynamics simulations indicate that fracture under uniaxial tension along the armchair direction is due to a break in the interlayer bond angles, while failure in the zig-zag direction is provoked by a break in both intra-layer angles and bonds. A modified Griffith criterion for the energy release-rate, and its dependence upon the strain-rates and crack orientations, indicated that the energy release-rate remains almost unchanged in the armchair direction but fluctuates wildly in the zig-zag direction. The strain-rate plays a negligible role in the energy release-rate. The geometrical factor in the Griffith criterion is almost constant when the crack orientation is less than 45°, regardless of the crack orientation and loading direction.[248]

Density functional tight-binding method studies of armchair and zig-zag nanotubes, subjected to uniaxial tensile strain, show that the deformation and failure of the nanotubes are very anisotropic. In the case of zig-zag nanotubes, 3 deformation stages are found. A primary linear elastic stage is associated with interactions between neighbouring puckers. This is followed by a bond-rotation stage in which the puckered configuration is smoothed-out by bond-rotation. In the third stage, bond-elongation occurs in which the P-P bonds are directly stretched up to the maximum limit and failure is initiated by rupture of the most highly-stretched bonds, which are located in specific rows. In the case of armchair nanotubes, the applied stress stretches the bonds up to the limit and causes failure. For both armchair and zig-zag nanotubes, the failure strain and stress are sensitive to the tube diameter while the Young's modulus, flexural rigidity, and Poisson ratio are

Materials Research Forum LLC

https://doi.org/10.21741/9781644903650

relatively insensitive.[249] In armchair nanotubes, the failure strain decreases from 0.40 to 0.25 and the failure stress increases from 13 to 21GPa as the tube diameter increases from 13.3 to 32.8Å. In zig-zag nanotubes, the failure strain decreases from 0.66 to 0.55 and the failure stress increases from 4 to 9GPa when the tube diameter increases from 13.2 to 31.1Å. The Young's modulus, flexural rigidity, radial and thickness Poisson ratios are 114.2GPa, 0.019eVnm^2, 0.47 and 0.11 for armchair nanotubes and 49.2GPa, 0.071eVnm^2, 0.07 and 0.21 for zig-zag nanotubes, respectively.[250]

In situ transmission electron microscopy and *in situ* micromanipulation have been combined in order to visualize directly the nanomechanical behaviour of individual nanoflakes. An easy liquid-phase shear exfoliation method can produce monolayer and few-layer nanoflakes in organic solvents, by using just a domestic blender. The nanoflakes can be easily bent, scrolled and stretched; demonstrating remarkable mechanical flexibility, without fracturing. An out-of-plane plate-like bending mechanism and an in-plane tensile strain of up to 34% are observed.[251]

Two-dimensional materials tend to crumple according to the Mermin-Wagner theorem and the resultant ripple deformation can markedly affect the electronic properties. First-principles calculations of a highly-anisotropic ripple pattern, where compression-induced ripple deformation occurred only along the zig-zag direction at strains of up to 10%, showed that this direction-selective ripple deformation arose from its puckered structure, with its coupled hinge-like bonding configurations, and from the resultant anisotropic Poisson ratio.[252] Phosphorene exhibits considerable structural flexibility along the armchair direction, permitting it to support high curvatures. Semiconduction and a direct band-gap are retained during buckling along the armchair direction. The band-gap decreases, and transforms into an indirect band-gap, upon buckling along the zig-zag direction. This structural flexibility facilitates the fabrication of complex shapes.[253]

First-principles studies of monolayer and bilayer phosphorene show that, under sufficiently high isotropic tensile strains, the monolayer loses its puckered structure and transforms into a flat hexagonal plane. Two inner sub-layers of bilayer phosphorene can be bonded due to its interlayer distance contraction. Under uniaxial tensile straining along a zig-zag direction, the pucker distance of each layer in bilayer phosphorene can exhibit a negative Poisson ratio. The properties of monolayer and bilayer phosphorene depend upon the magnitude and direction of the applied strain. Their band-gaps decrease more rapidly under isotropic compressive strain than under uniaxial strain. Their direct-indirect band-gap transitions occur at higher isotropic tensile strains than that under uniaxial strain. Under isotropic compressive strain, bilayer phosphorene undergoes a transition from a direct-gap semiconductor to a metal. On the other hand, monolayer phosphorene initially undergoes a direct-indirect transition, and then changes into a metal. Under isotropic tensile straining, both the bilayer and monolayer phosphorene undergo a direct-indirect transition and finally change into a metal.[254] When an uniaxial stress is applied along an arbitrary direction, there is an enhancement of the inherent anisotropy.[255]

First-principles calculations show that phosphorene exhibits a markedly anisotropic elasticity with an anisotropy factor of 15.5. Sheet samples exhibit a high tensile ductility along the armchair direction, characterized by an elastic strain limit of 0.31. In the zig-zag direction, the critical strain is dictated by the phonon instability and an in-plane soft mode occurs beyond a strain of 0.22. Under uniaxial strain, the band-gaps can be continuously modulated.

The edge elastic modulus and edge stress of zig-zag nanoribbons are about three times larger than those of armchair nanoribbons. The tensile edge stresses along zig-zag and armchair edges are able to provoke a distortion of free-standing nanoribbons.[256]

By using a density functional based potential and realistic boundary conditions, the flexibility of individual α, β and γ allotrope layers along an arbitrary bending direction was investigated. Correlation of the results with in-plane elasticity showed that an orthotropic plate with a direction-dependent thickness could describe the out-of-plane deformation of the α and γ allotropes.[257] First-principles calculations indicated that δ-phosphorene nanosheets are good auxetic materials, with a high negative Poisson ratio. The Young's modulus and Poisson ratio are anisotropic. The negative Poisson ratio along the grooved direction has values of up to -0.267. The auxetic nature of δ-phosphorene arises from its puckered structure and is insensitive to the number of layers, due to weak interlayer interactions.[258] The δ-phosphorene possesses a good flexibility because of its relatively low Young's modulus and high critical crack strain.

Mode-I stress intensity factors were estimated[259,260] for blue phosphorene and the crack-tip displacement field which was predicted by linear elastic fracture mechanics was used to determine the stress intensity factor (table 36). Atomic-scale finite-element methods and Stillinger-Weber potentials were used to simulate tensile tests. An empirical formula was derived which related the mode-I fracture toughness to the elastic modulus, the intrinsic tensile strength, the bond-length and the buckling height of atom-thick hexagonal material.

Table 36. Properties of phosphorene

Property	Value
lattice constant	3.326Å
bond length	2.27Å
buckling height	1.211Å
Van der Waals radius	1.8Å
E(armchair)	59.9N/m
E(zig-zag)	60.6N/m
Poisson ratio(armchair)	0.17
Poisson ratio(zig-zag)	0.17
maximum stress(armchair)	7.9N/m
maximum stress(zig-zag)	7.3N/m
strain(armchair)	22%
strain(zig-zag)	23.5%
K_I(armchair)	0.62MPa√m
K_I(zig-zag)	0.58MPa√m

First-principles calculations were used[261] to study the ideal tensile stress-strain relationships of the monoatomic monolayer 2-dimensional buckled honeycomb lattice. The ideal strengths and critical strains were investigated for uniaxial and equi-biaxial strains (table 37). It was concluded that the critical strain was governed by the simultaneous stretching and rotation of bonds.

Materials Research Forum LLC
https://doi.org/10.21741/9781644903650

Table 37. Ideal strength and Young's modulus of phosphorene

Property	Orientation	Value(N/m)
strength	armchair	7.06
strength	zig-zag	6.72
strength	biaxial	7.29
modulus	-	79.7

The mechanical properties of defected nanotubes under uniaxial tension were predicted[262] by performing molecular dynamics simulations. These showed that vacancies created local stress concentrations and thus markedly reduced the fracture strength and fracture strain. A 1% defect concentration was able to reduce the fracture strength and fracture strain by up to 50% and 66%, respectively. The impairment of properties depended upon the defect's location: a defect in the outer sub-layer had a greater effect than did one which was located in the inner layer; especially for nanotubes of small diameter. The temperature strongly affected the properties of both defect-free and defected nanotubes. As the temperature was increased from 0K to 400K, the fracture strength and fracture strain of nanotubes with a 1% defect concentration were reduced by 71% and 61%, respectively.

Multi-layer assemblies in which flakes were piled up in a staggered fashion could exhibit an exceptional toughness. Molecular dynamics simulations were used to study[263] the dependence of the mechanical properties upon the overlap distance of adjacent layers and the number of flakes per layer. When the flake number was 1, there was a transition in fracture-pattern upon increasing the overlap distance: from ductile failure controlled by interfacial friction to brittle failure dominated by the breaking of covalent bonds within the flakes. The failure pattern could be tuned by changing the number of flakes in each layer. The ultimate strength obeyed a power-law with an exponent -0.5 with regard to the number of flakes. The optimum number of flakes in each layer was 2 at 1K in order to ensure both a high toughness and strength. The relationship between mechanical properties and overlap distance could be explained in terms of a shear-lag model. Increasing the temperature could change the fracture pattern from ductile to brittle. The optimum number of flakes depended very much upon the temperature with regard to producing a high strength and toughness.

Molecular dynamics simulations indicated that fracture under uniaxial tension along the armchair direction could be attributed to a break in interlayer bond angles[264], while failure in the zig-zag direction is initiated by a break in intra-layer angles and bonds. A modified Griffith criterion was based upon the energy release-rate and its dependence upon the crack strain-rate and orientation. Simulation also indicated that the energy release-rate is almost unchanged in the armchair direction, but fluctuates intensively in the zig-zag direction. The strain-rate had little effect upon the energy release-rate. The geometrical factor in the Griffith criterion was essentially constant when the crack orientation was less than 45°, regardless of the crack orientation and loading direction.

An isomer called green phosphorus was theoretically predicted which should exhibit a similar interlayer interaction to that of black phosphorus[265]. This suggested that individual layers could be mechanically exfoliated in order to produce 2-dimensional layers of the so-called green phosphorene. The properties of its nanoribbons along the armchair and zig-zag directions were predicted for widths of up to 57Å by using density functional theory. The ribbons were notably anisotropic, with the Young's modulus ranging from 10 to 35GPa for armchair nanoribbons and from 160 to 170GPa for zig-zag nanoribbons. These values were of the same order of magnitude as those for 2-dimensional sheets.

The deformation of black phosphorene nanoflakes having a thickness of 6 to 22nm (11 to 43 layers) was studied[266] by means of atomic force microscopic nano-indentation. The modulus increased as the number of layers decreased, according to:

$$E = 53.31(1 + 3.30e^{-0.13n})$$

and the modulus for a monolayer (n = 1) was 214N/m.

Reactive molecular dynamics simulations were made[267] of the room-temperature mechanical properties of β-, γ-, δ-, θ- and α-phosphorene, a sheet with 4-8 membered rings and 2 sheets with 5-7 membered rings. Various anisotropies in the mechanical properties depended upon the crystal structures. The Young's modulus and tensile strength ranged from 7.3% to 25% and from 8.6% to 22% of those of graphene, respectively. In the early stages of fracture, eye-like cracks formed via local bond-breaking which was perpendicular to the tensile direction in hexagonal and 4-8 sheets. Complete fracture involved straight cracks in hexagonal sheets under tension in the zig-zag direction and under tension along the square edge direction in 4-8 sheet. Crack meandering and branching occurred under tension in α-, β- and γ-phosphorene in the armchair direction and along the square diagonal direction in 4-8 sheet. Under uniaxial tension of sheets with 5-7 atom rings, 12- and 10-membered rings formed via the merging of two neighbouring heptagons, and a heptagon and its neighbouring pentagon, respectively. The 12- and 10-membered rings then coalesced, causing failure of the sheets.

The properties of nanoribbons were investigated[268] by using density functional theory, showing that their ideal strength was lower than that of 2-dimensional phosphorene. The Young's modulus of armchair nanoribbon exhibited a notable size-effect because of edge relaxations. Analysis of stress-strain relationships indicated that chemisorbed oxygen greatly reduced the ideal strength and Young's modulus of 2-dimensional phosphorene oxide in the zig-zag direction, especially at high oxidation-ratios.

Density functional tight-binding methods were used to study the deformation and failure of armchair and zig-zag nanotubes under uniaxial tensile straining[269]. In the case of the zig-zag group, three deformation-stages were identified. There was a primary linear elastic stage which was associated with interactions between neighbouring puckers, that was followed by a bond-rotation stage in which the puckered configuration was smoothed by bond-rotation. There was then a bond-elongation stage in which the P-P bonds were stretched to the limit and failure was triggered by breaking of the most highly stretched bonds. In the case of armchair samples, the applied force stretched the bonds up to the limit and led to final failure. In both samples, the failure strain and failure stress were affected by the nanotube diameter. The Young's modulus, the flexural rigidity, the radial Poisson ratio and the thickness Poisson ratio were relatively unaffected by the nanotube diameter. In armchair nanotubes, the failure strain decreased from 0.40 to 0.25 and the failure stress increased from 13GPa to 21GPa as the tube diameter increased from 13.3Å to 32.8Å. In zig-zag nanotubes, the failure strain decreased from 0.66 to 0.55, and the failure stress increased from 4GPa to 9GPa as the tube diameter increased from 13.2Å to 31.1Å. The Young's modulus, radial and thickness Poisson ratios were 114.2GPa, 0.47 and 0.11 for armchair nanotubes, and 49.2GPa, 0.07 and 0.21 zig-zag nanotubes, respectively.

Molecular dynamics finite element methods and the Stillinger-Weber potential were used[270] to simulate uniaxial tensile testing of pristine and defected black phosphorene sheets, revealing that the effect of defects upon the Young's modulus was negligible. On the other hand, defects could reduce the fracture stress and fracture strain by up to 53% and 69%, respectively.

The properties of phosphorene oxides were investigated[271] by using density functional theory. These showed that, as the oxygen coverage approached 50%, negative Poisson ratios appeared in the in-plane principal directions. These arose from an asymmetrical atomic structure which resulted from oxygen adsorption. The overall structure of phosphorene oxide was weakened as compared to that of phosphorene.

A 2-dimensional violet phosphorene compound, composed of crossed sub-nano rods, exhibited an extremely high deformation-resistance[272]. The 2-dimensional Young's modulus of monolayer violet phosphorene was found to be 1512N/m by means of nano-indentation and atomic force microscopy. The 2-dimensional Young's modulus of monolayer violet phosphorene was 4.4 times higher than that of graphene. The extremely high bending stiffness was attributed to its staggered crossed structure.

Density functional calculations estimated[273] the in-plane Youngs's modulus to be 102.45N/m and 23.43N/m along zig-zag and armchair directions, respectively; a marked anisotropy. The material was very flexible and could support a tensile strain up to 35% along the zig-zag direction and 70% along the armchair direction. It was predicted that the band-gap could be varied from 0 to 1.12eV by imposing compressive and tensile strains.

On the basis of first-principles calculations, δ-phosphorene nanosheets were expected[274] to be good auxetic materials with a high Poisson ratio. The Young's modulus and Poisson ratio were anisotropic. The Poisson ratio along the grooved direction could be up to - 0.267. The auxetic effect arose from its puckered structure and was insensitive to the number of layers, due to weak interlayer interaction. The material also exhibited a good flexibility, due to its relatively small Young's modulus and high critical crack strain.

The properties of armchair nanotubes were predicted[275] by using combined density functional theory and finite-element methods. The Young's modulus of the nanotubes was closely related to the nanotube radius. An increase in the aspect-ratio of the nanotubes corresponded to an increase in the Young's modulus; apart from a marked exception at low aspect-ratios, where the effect upon elastic properties was negligible.

First-principles calculations were used[276] to study the response to external strain of material which contained a substitutional amount of impurity. The formation-energy of defect structures could increase or decrease with bi-axial tensile or compressive strains, depending upon the atomic radius of the impurity atom; larger or smaller than that of the host atoms. Strain-maps indicated that this behaviour was related to the local compressive or tensile local strains which were produced by the impurities, given that they interfered with the external strain. The change in the defect-formation energy was related to the change in elastic modulus upon introducing an impurity.

A finite-element model was used to investigate[277] the properties of zig-zag nanosheets which were formed from beam elements. The properties of the beam elements were functions of the force-constants. The model could predict the mechanical properties of monolayer zig-zag nanosheets under tensile and compressive loadings.

The mechanical properties of black phosphorene and blue phosphorene were predicted[278] by means of molecular dynamics simulations. The blue material had isotropic mechanical properties, unlike the anisotropy of the black material. The differing properties were attributed to their different structures. The temperature had little effect upon the Young's modulus of either material, but the fracture stress decreased linearly with increasing temperature. In general, the reduction in fracture strain was rapid at lower temperatures.

Density functional tight-binding simulations were used[279] to study the deformation and failure behaviour of nanoribbons which had edges which were passivated by hydrogen atoms. Edge-relaxation occurred mainly within the 3 atomic rows nearest to the passivated edges, and was in the form of alterations in the bond-length and bond-angle. Edge-passivation could reduce edge-relaxation and reduce edge-stresses and elastic moduli. In armchair nanoribbons under tension, 3 deformation-stages could be identified. One was linear elastic, and involved bond-rotation and bond-stretching. At a strain of about 0.5, the intra-row bonds fractured in a double-row manner within the interior of the material, leading to fragmentation. In zig-zag nanoribbons under tension, inter-row bonds were directly stretched. At a strain of about 0.18, a fraction of the inter-row bonds broke at random, leading to failure. Edge-passivation increased the failure strain of zig-zag material by about 80% and the strength by about 26%. The corresponding increases for armchair material were 2% and 5%, respectively.

Molecular dynamics simulations were used to study[280] the effect of the grain size upon the mechanical behaviour of polycrystalline sheets. Unlike other 2-dimensional materials having a planar structure, these polycrystalline phosphorene sheets were almost linearly elastic, due to their high bending stiffness and intrinsic buckled structure. The increase in stiffness of the polycrystalline material, associated with an increase in the grain size from 2 to 12nm was 15.9%. This was much smaller than that for other 2-dimensional materials having a planar crystalline structure. The insensitivity was attributed to the small difference between the elastic modulus of the crystalline and amorphous phases. The strength reduction obeyed a logarithmic dependence upon the grain size, and this was explained in terms of dislocation pile-up theory.

Molecular dynamics simulations were used[281] to study the mechanical properties under shear loading, showing that the shear modulus was about 22GPa for both the armchair and zig-zag directions. The fracture strength and ultimate strain were markedly reduced by a greater thermal vibration of atoms at higher temperatures.

Defects

Due to their low formation energies, it is very easy to create atomic defects in phosphorene during fabrication. The defect formation energy in phosphorene is generally much lower than that in graphene, and the defect configuration strongly affects the electronic structure. The band-gap vanishes in single-vacancy layers, but reappears in divacancy layers. A single vacancy behaves like a p-type impurity with regard to the transport properties. Density functional theory, combined with semi-empirical van der Waals dispersion corrections, shows that a vacancy or a self-interstitial defect is more stable in the outer layer than in the inner layer of defective few-layer phosphorene The formation energy and transition energy of a vacancy or a self-interstitial phosphorus defect decrease with increasing film thickness, due mainly to an upward shift of the host valence band maximum relative to the vacuum level.

Two possible monovacancies and 14 possible divacancies have been identified in phosphorene. Monovacancy-(5|9), with a pair of pentagon-nonagons, is the ground state of monovacancies and divacancy-(5|8|5), with a pentagon-octagon-pentagon structure, is the most stable divacancy. The 14 divacancies can be further sub-divided into four basic types, depending upon their topological structures, and transform between these configurations via bond-rotation. The diffusion of monovacancy-(5|9) prefers to migrate along the zig-zag direction in the same half-layer. The introduction of monovacancies and divacancies affects the vibrational properties and introduces localized defect modes. These can be used to distinguish between the various vacancy structures.

Direct observation of the monovacancies, with their high mobility and anisotropic motion, is possible only at about 70K, where thermal activity is largely suppressed. Because of the low diffusion barrier (0.3eV), the motion of a monovacancy at room temperature is 16 orders of magnitude faster than it is in graphene. The intrinsic strain which is associated with the vacancies extends far along the zig-zag direction, but falls off rapidly along the armchair direction. The motion of divacancies, via multiple dissociation-recombination processes involving monovacancies, exhibits novel features due to the small energy-cost (about 1.05eV) of splitting a divacancy into two monovacancies.[282] Uniaxial tensile strain, applied along the zig-zag direction, can promote the motion of monovacancies while tensile strain, applied along the armchair direction, had the opposite effect.

Vacancies introduce local stress concentrations and provoke early bond-breaking, leading to a marked degradation of the mechanical properties. A 2% concentration of randomly distributed monovacancies can reduce the fracture strength by about 40%. An increase in temperature from 10 to 400K further degrades the fracture strength by about 60%. The fracture strength of defective phosphorene is affected by the defect distribution.[283] When the defects are arranged in a line, the reduction in fracture strength depends markedly upon the tilt-angle and the loading direction. Divacancies cause a greater reduction in fracture strength than do monovacancies, when loading is applied in the armchair direction.

First-principles calculations of the stress-strain relationship of phosphorene which contains vacancies indicate that, due to the presence of the vacancies, properties such as the ideal strength and Young's modulus are greatly degraded in the zig-zag direction.[284] Tensile strain produces a marked reduction in the band-gap, and lithium-diffusion through the defected material is more energetically favourable than in pristine phosphorene. This effect is further increased by applying biaxial tension.

Molecular dynamics simulations show that the Young's modulus, shear modulus, specific heat capacity and coefficient of thermal expansion of armchair and zig-zag sheets decrease with increasing vacancy defect ratio.

During the synthesis of 2-dimensional materials, line defects are often generated at the grain boundaries between domains of differing orientation. Twelve types of line-defect structure in monocrystalline phosphorene have been studied using first-principles calculations. These line defects usually formed via the migration and aggregation of intrinsic point defects such as Stone-Wales, monovacancy or divacancy. The migration of point defects in phosphorene is anisotropic. The lowest migration energy barriers are 1.39 and 2.58eV for Stone-Wales defects in the zig-zag and armchair directions, respectively. The lowest migration-energy barriers are 0.40 and 0.49eV for monovacancies in the zig-zag and armchair directions, respectively.[285] The aggregation of point defects into lines is energetically favoured over separate point defects. The axis of line defects is direction-selective, depending upon the type of point defect of which it is composed.

Blue phosphorene can be used to prepare small nanotubes, and it is possible to create low-energy nanotubes via the inclusion of defect lines. Such defect-induced faceted nanotubes, made using five different types of defect line, exhibited negligible bending stresses. This led to a reduction in the formation energy with respect to round nanotubes. Armchair faceted nanotubes had similar formation energies to those of multiphase faceted nanotubes, but offered a wider range of possible structures. Zig-zag faceted nanotubes had lower formation energies than did round tubes or multiphase faceted nanotubes.

High-angle tilt boundaries with a higher density of (5|7) defect pairs oriented along the armchair direction are stronger than are low-angle tilt boundaries with a lower defect density. High-angle boundaries with a higher density of (4|8) defect pairs oriented along the zig-zag direction are similarly stronger than are low-angle boundaries having a lower defect density. Failure is due to the rupture of the most pre-strained bonds in the heptagons of the (5|7) defect pair, or octagons of the (4|8) pair. High-angle grain boundaries are more able to accommodate pre-strained bonds in heptagon and octagon defects, leading to higher failure stresses and strains.[286] These results cannot be explained in terms of a Griffith-type fracture-mechanics criterion because it does not take account of bond pre-stretching. Such anomalous mechanical and failure characteristics of phosphorene tilt grain boundaries are shared by graphene. It was suggested that they may be universal features of 2-dimensional materials.

Bismuthene

First-principles phonon and finite-temperature molecular dynamics calculations, incorporating spin-orbit coupling, showed that free-standing single-layer buckled-honeycomb and asymmetrical-washboard bismuthene structures are stable at high temperatures. Spin-orbit coupling is found to be essential to determining lattice constants, phonon frequencies, band-gaps and cohesion. The phonons of 3-dimensional hexagonal crystals, as well as those of single-layer bismuthene, are softened by spin orbit coupling. In going from the 3-dimensional hexagonal crystal to free-standing single-layer structures, the 2-dimensional hexagonal lattice is compressed and the semimetal is transformed into a semiconductor by the confinement effect.[287] On the other hand, by going from single-layer to bilayer bismuthene, the lattice is slightly expanded and the fundamental band-gaps are narrowed. Bismuth exhibits the highest spin-orbit coupling strength among the main group elements. The quantum anomalous Hall state is predicted to occur in half-hydrogenated bismuth honeycomb monolayers, Bi_2H. Halogenated films, Bi_2X (X = F, Cl, Br), have a metallic nature which blocks quantum anomalous Hall effects.[288] A double cantilever beam fracture technique can be used to transfer epitaxial bismuth thin films, grown on silicon (111), to silicon strips coated with epoxy. The transferred films retain electrical, optical and structural properties which are comparable to those of as-grown epitaxial films. The adhesion energy at the bismuth/silicon interface is about $1J/m^2$, and comparable to that of exfoliated and wet-transferred graphene.[289]

Mechanical Properties

Mode-I stress intensity factors were estimated[290,291] for bismuthene and the crack-tip displacement field which was predicted by linear elastic fracture mechanics was used to determine the stress intensity factor (table 38). Atomic-scale finite-element methods and Stillinger-Weber potentials were used to simulate tensile tests. An empirical formula was derived which related the mode-I fracture toughness to the elastic modulus, the intrinsic tensile strength, the bond-length and the buckling height of atom-thick hexagonal material.

Table 38. Properties of bismuthene

Property	Value
lattice constant	4.340Å
bond length	3.045Å
buckling height	1.73Å
Van der Waals radius	2.07Å
E(armchair)	26.9N/m
E(zig-zag)	27.3N/m
Poisson ratio(armchair)	0.24
Poisson ratio(zig-zag)	0.24
maximum stress(armchair)	5.1N/m
maximum stress(zig-zag)	4.6N/m
strain(armchair)	27.8%
strain(zig-zag)	29.8%
K_I(armchair)	0.37MPa√m
K_I(zig-zag)	0.34MPa√m

Density functional theory was used to investigate[292] the properties of pristine material and of adsorbed bismuthene with adsorbed hydrogen, fluorine, chlorine or bromine atoms (table 39). The elastic moduli of pristine and adsorbed nanosheets under uniaxial loading, and their bulk moduli under biaxial strain, were calculated. Adsorption led to a decrease in the elastic and bulk moduli. The effect of adsorption upon the elastic modulus of the bismuthene (table 40) was greater than its effect upon the bulk modulus (table 41). The yield strain of bismuthene nanosheet under uniaxial loading was decreased by the adsorption.

Table 39. Characteristics of adsorbed bismuthene nanosheets

Material	Lattice Constant(Å)	Adsorbed Bond Length(Å)	Buckling(Å)
Bi	4.4788	-	1.775989
BiH	5.6959	1.845	0.124089
BiF	5.5414	2.134	0.353395
BiCl	5.6547	2.551	0.199214
BiBr	5.6856	2.709	0.153003

Table 40. Young's moduli of pristine and halogen-adsorbed
bismuthene

Material	$E_{longitudinal}$(N/m)	$E_{transverse}$(N/m)
Bi	26.25	26.25
BiH	12.84	12.84
BiF	11.58	10.36
BiCl	9.96	9.95
BiBr	10.32	10.30

Table 41. Bulk moduli of pristine and
halogen-adsorbed bismuthene

Material	B(N/m)
Bi	16.70
BiH	20.88
BiF	12.18
BiCl	13.52
BiBr	13.80

The mechanical properties of monolayer bismuthene under uniaxial and biaxial strains were investigated[293] by using a first-principles approach. It remained stable under a wide range of strains, and the Young's modulus varied in a consistent manner to the Poisson ratio. A negative Poisson ratio existed under appreciable tensile straining.

By using the Stillinger-Weber potential, molecular dynamics simulations were made[294] of the mechanical strength of monolayer β-bismuthene. Analysis of the effect of temperature upon the tensile properties along the armchair and zig-zag directions of nanosheets showed that an increase in temperature led to an appreciable decrease in the properties. The fracture resistance was higher under zig-zag loading, while the armchair direction led to higher elasticity. Increases in vacancy concentration and crack length markedly reduced the fracture stress and strain. For all of the conditions, the β-bismuthene exhibited a clear chirality effect under tensile loading. In the case of pre-cracked samples, armchair-directed cracks were associated with a higher fracture resistance than were zig-zag directed cracks. Branching occurred during the propagation of armchair cracks. The crack propagated perpendicularly to the loading direction for zig-zag cracks.

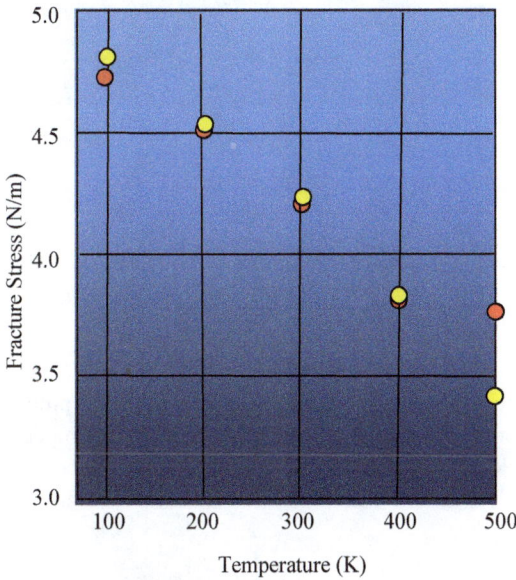

Figure 13. Fracture stress of β-bismuthene, as a function of temperature under tension in the armchair (yellow) and zig-zag (red) directions

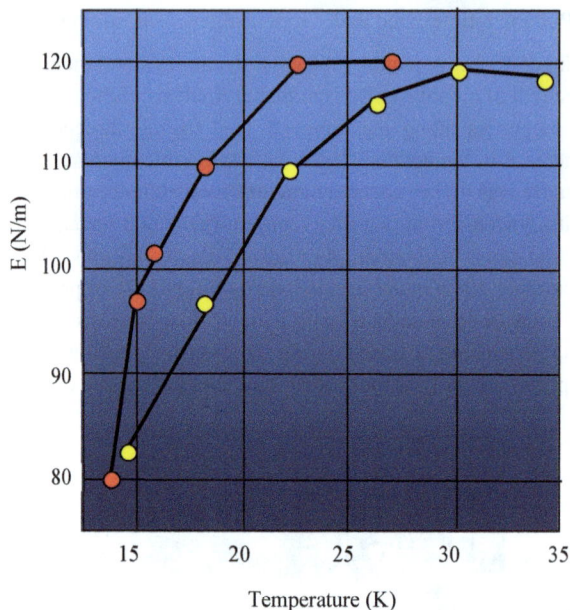

Figure 14. Young's modulus of β-bismuthene as a function of temperature under tension in the armchair (yellow) and zig-zag (red) directions

Density functional theory was used to determine the mechanical properties of monolayer bismuthene in the presence of an electric field[295]. The Young's modulus was insensitive to small electric fields, although a high electric field could affect it. A sufficiently high electric field could decrease the yield strain. The Bi-Bi bond-length was insensitive to an external electric field.

*Table 42. Longitudinal Young's modulus of
bismuthene nanosheet in an external electric field*

Electric Field(V/Å)	$E_{longitudinal}$(N/m)
0	26.2519
0.1	26.2386
0.2	26.2415
0.26	26.2475
0.32	26.2792
0.38	26.2863
0.42	26.2674
0.46	25.9995

*Table 43. Transverse Young's modulus of bismuthene
nanosheet in an external electric field*

Electric Field(V/Å)	$E_{transverse}$(N/m)
0	26.2504
0.1	26.2371
0.2	26.2305
0.26	26.2175
0.32	26.2894
0.38	26.5986
0.42	26.6376
0.46	26.6499

Table 44. Bulk modulus of bismuthene nanosheet in an
external electric field

Electric Field(V/Å)	B(N/m)
0	16.6984
0.1	16.6981
0.2	16.7049
0.26	16.7113
0.32	16.7049
0.38	16.6920
0.42	16.6984
0.46	16.6920

Borophene

Pure boron is quite similar to carbon, and forms simple planar molecules and cage-like fullerenes. Theory predicts that 2-dimensional boron sheets will take up an atomic configuration which is similar to that of boron atomic clusters. Atomically thin crystalline 2-dimensional boron sheets can be prepared on silver surfaces under ultra-high vacuum conditions. Detailed characterization reveals structures which recall fused boron clusters, with multiple scales of anisotropic out-of-plane buckling. Unlike bulk boron, borophene exhibits metallic characteristics which are consistent with a highly anisotropic 2-dimensional metal.[296] Elemental boron, being electron-deficient, cannot form graphene-like structures but triangular lattices containing hexagonal vacancies have been predicted to be stable.

Structure

Experimental and computational studies show that the B_{36} cluster has a planar C_{6v} structure with a central hexagonal hole, thus providing experimental evidence for the viability of atomically-thin boron sheets. Neutral B_{36} is the smallest boron cluster to have six-fold symmetry, plus a perfect hexagonal vacancy.[297] A boron cluster with a double-hexagonal vacancy is a more flexible structural motif. The photo-electron spectrum of B_{35} has a simple pattern, having some similarity to that of B_{36}. Global-minimum searches indicate that B_{35} possesses a planar hexagonal structure which is similar to that of B_{36}, apart from a missing interior boron atom that creates the double-hexagonal vacancy. Closed-shell B_{35} exhibits triple π aromaticity, with 11 delocalized π-bonds; analogous to benzo(g,h,i)perylene ($C_{22}H_{12}$). The B_{35} cluster can be used to build atomically-thin boron sheets having various hexagonal hole densities.[298] Combined photo-electron spectroscopic and theoretical studies also indicate the existence of inherently chiral boron clusters, $[B_{30}]^-$. The most stable structure of $[B_{30}]^-$ is quasi-planar, with an hexagonal hole. Two enantiomers, arising from differing positions of the hexagonal hole, are degenerate. Both of them should co-exist experimentally because they have identical electronic structures and give rise to identical simulated photo-electron spectra.[299]

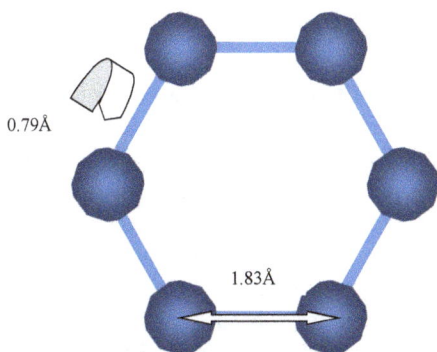

Figure 15. Basic borophene hexagon, showing B-B
bond-length, bond-angle and buckling

First-principles calculations of monolayer 8-Pmmn borophene attribute amazing electronic properties to the 2-dimensional material. On the basis of Born effective charge analysis, 8-Pmmn borophene is deduced to be the first single-element based monolayered material to exhibit 2 sub-lattices having substantial ionic features. A tight-binding model confirms that the crystal hosts massless Dirac fermions and that the Dirac points are protected by symmetry.[300] When strain is introduced into the model, it introduces a pseudomagnetic field vector potential and a scalar potential. The observed Dirac cones are formed by the p_z orbitals of one of the inequivalent sub-lattices, which are composed of just 4 atoms, yielding an underlying hexagonal network which is topologically equivalent to that of distorted graphene. This offers the possibility of converting metallic 8-Pmmn borophene into an indirect band-gap semiconductor by applying an external shear stress.[301] Stability of the strained structures is supported by phonon frequency analysis. The Dirac cones are sensitive to the formation of vacancies only in the inequivalent sub-lattice which is active at the Fermi level.

There was some experimental evidence that borophene which formed on a Ag(111) surface consisted of a buckled triangular lattice without vacancies. Calculation suggested a nucleation mechanism for boron clusters and emphasized boron-silver interactions during the growth of borophene, including the structural evolution of triangular fragments having various profiles and vacancy distributions. When compared with a triangular lattice without vacancies, it was confirmed that the sheet which was energetically favoured during nucleation and growth was one which contained some 16% of vacancies in a stripe pattern.[302]

Mechanical Properties

The mechanical properties of bilayer material were recently predicted[303] by using first-principles and anisotropic Migdal-Eliashberg methods. The coexistence of strong covalent bonds and de-localized metallic bonds imparted good mechanical properties, such as a maximum 2-dimensional Young's modulus of some 570N/m. The superconducting critical temperature could also be increased to about 46K by an applied strain; currently the highest known value for borophenes or 2-dimensional elemental materials. Previous molecular dynamics results had shown that the Young's moduli of armchair and zig-zag borophene nanoribbons were 1.093 and 0.978TPa, respectively, and that the corresponding elastic strains were about 15.30 and 22.03%, with distinct ductile and brittle fractures, respectively. When the nanoribbons were grafted with amine groups (-NH$_2$), the moduli of the armchair and zig-zag nanoribbons increased to 1.125 and 1.016TPa, respectively. The ultimate elastic strain for the former meanwhile increased to 18.23%, but that of armchair nanoribbons was slightly reduced to 21.12%. The fracture modes remained the same. Structural deformation following exposure to a temperature of 1500K showed that there was little resultant change between the armchair and zig-zag nanoribbons, but the structural deformation of the grafted nanoribbons was clearly less than that for non-grafted nanoribbons. The overall conclusion was that the grafting of amine groups could increase the Young's modulus, lengthen the elastic-strain range, reduce in-plane elastic anisotropy, increase crack resistance, increase the capacity to resist deformation at high temperatures, reduce thermal-expansion anisotropy and improve structural stability.[304]

Two-dimensional sheets having rectangular structures have been successfully grown onto monocrystalline Ag(111) substrates. First-principles calculations predicted that monolayer borophene could withstand stresses of up to 20.26 and 12.98N/m in the a- and b-directions, respectively, but that the critical strain was small. In the a-direction, the critical value was 8%, and this seemed to be lowest among all of the studied 2-dimensional materials.[305] A tensile strain which was applied in the b-direction enhanced the bucking-height and resulted in a negative out-of-plane Poisson ratio, thus leading a sheet to possess a superior mechanical flexibility along the b-direction. Molecular dynamics simulations of the properties of 5 types of film, having vacancy ratios ranging from 0.1 to 0.15, showed that the Young's modulus and tensile strength decreased with increasing temperature. Sheets (table 45) exhibited an anisotropic mechanical response due to the differing arrangements of atoms along the armchair and zig-zag directions. At room temperature, the 2-dimensional Young's modulus and fracture stress ranged from 63 to 136N/m and from 12 to 19N/m, respectively. The strains at the ultimate tensile strength (table 46) ranged from 9 to 14%, from 11 to 19% and from 10 to 16% at 1, 300 and 600K, respectively[306], reflecting the marked stiffness of the 2-dimensional material.

Table 45. Dimensions of borophene sheets used in simulations

Type	Length$_{zig-zag}$(Å)	Length$_{armchair}$(Å)	Number of Atoms
1/9	143.86	149.03	5600
1/8(A)	136.91	137.12	4830
1/8(B)	136.89	137.16	4830
2/15	136.90	137.16	4784
4/27	144.19	143.53	5152

Table 46. Mechanical properties of boronphene sheets

Type	T(K)	Direction	E(N/m)	Strength(N/m)	Strain(%)	Toughness(J/m^2)
1/9	1	zig-zag	249	33.1	8.66	1.27
1/9	1	armchair	284	38.1	10.0	1.86
1/9	300	zig-zag	128	14.7	15.41	1.21
1/9	300	armchair	136	14.1	11.84	0.75
1/9	600	zig-zag	109	11.2	9.9	0.65
1/9	600	armchair	115	10.4	10.3	0.51
1/8(A)	1	zig-zag	288	32.3	10.5	1.69

1/8(A)	1	armchair	284	29.6	9.18	1.31
1/8(A)	300	zig-zag	123	12.3	11.0	0.65
1/8(A)	300	armchair	126	15.2	14.16	1.00
1/8(A)	600	zig-zag	97	10.2	11.2	0.59
1/8(A)	600	armchair	114	10.8	10.7	0.62
1/8(B)	1	zig-zag	189	13.4	13.7	1.16
1/8(B)	1	armchair	290	36.9	11.7	1.69
1/8(B)	300	zig-zag	63	12.0	17.3	0.98
1/8(B)	300	armchair	126	18.4	16.5	1.62
1/8(B)	600	zig-zag	60	9.1	14.1	0.66
1/8(B)	600	armchair	99	13.1	13.75	0.97
2/15	1	zig-zag	280	33.4	11.8	2.07
2/15	1	armchair	228	25.5	9.85	1.19
2/15	300	zig-zag	119	19.4	18.6	1.81
2/15	300	armchair	116	15.0	14.3	1.03
2/15	600	zig-zag	100	11.9	13.1	0.81
2/15	600	armchair	82	9.7	15.1	0.84
4/27	1	zig-zag	180	22.2	12.5	1.38
4/27	1	armchair	192	24.8	13.3	1.55
4/27	300	zig-zag	95	14.0	18.0	1.36
4/27	300	armchair	96	13.1	15.0	1.03
4/27	600	zig-zag	87	11.0	15.1	0.96
4/27	600	armchair	88	9.5	15.6	0.95

Single-layer borophene has a buckled structure, with atomic ridges. First-principles density functional theory calculations showed that the 2-dimensional borophene is very stretchable, with a marked anisotropy The strain-to-failure in the direction along the atomic ridges is nearly twice as high as that across the atomic ridges. Straining-induced flattening, and subsequent stretching of the flattened structure, account for the large strain-to-failure in tension along the atomic ridges. Biaxial tension increases the strength across the atomic ridges, but decreases the failure-strain along the atomic ridges.[307] When bilayer borophene is stretched along the cross-plane direction, the strength and failure strains are much higher than those of bilayer graphene. This is due to the very strong interlayer chemical bonding. Two-dimensional materials tend to be mechanically flexible yet planar, especially when adhering to metal substrates. First-principles calculations showed that periodic nanoscale 1-dimensional undulations can be preferred in borophene located on reconstructed Ag(111). Such a wavy configuration was more stable, than its planar form on flat Ag(111), due to the highly anisotropic bending flexibility of borophene. Atomic-scale ultra-high vacuum scanning tunnelling microscopy characterization of borophene, grown onto Ag(111), revealed such undulations. These agreed with theory in terms of topography, wavelength, moiré pattern and prevalence of vacancy defects.[308] Although the lattice was coherent within a borophene island, undulations which had nucleated from different sides of the island formed a distinctive domain boundary when they were laterally misaligned. This structural model suggested that the transfer of undulated borophene to an elastomeric substrate would permit high levels of stretchability and compressibility.

First-principles calculations, based upon density functional theory, showed that monolayer 8-Pmmn borophene could withstand stresses of up to 21.59 and 27.79N/m in the a- and b-directions, respectively. The ultimate tensile strain along the a-direction was about 0.155 and that along the b-direction was 0.16; higher than that of stripe borophene.[309] The phonon dispersion under applied strain indicated that mechanical failure was likely to originate from elastic instability. Overall, 8-Pmmn borophene is a hard but flexible material.

The mechanical properties of 2-Pmmn borophene were deduced[310] by means of first-principles calculations. The Young's moduli were 236 and 89GPa in the armchair and zig-zag directions, respectively. It was electroconductive in the armchair direction but had a band-gap in the zig-zag direction. The resistance decreased with increasing strain, and this was attributed to expansion of the buckled structure.

Single-layer material was shown[311], using molecular dynamics simulations, to exhibit highly anisotropic mechanical properties. The Young's modulus and fracture strength in the zig-zag direction were much lower than those in the armchair direction. The simulated fracture strength and fracture strain at 10K agreed well with those predicted by first-principles calculations. The fracture properties were very sensitive to temperature and, as the temperature increased from 10 to 700K, the fracture strength and fracture strain decreased by about 50% and 60%, respectively. The fracture properties exhibited however a relatively weak dependence upon the strain-rate. When the latter was changed from 0.00001 to 0.01ps, the fracture strength and fracture strain increased by only 6.8 to 8.6%, and by 11 to 15%, respectively. The effects of temperature and strain-rate upon the fracture strength were explained in terms of the kinetic theory of fracture.

The behaviour of few-layer η-borophene at 10 to 800K for various strain-rates was studied[312] by means of molecular dynamics simulations which were based upon the Stillinger-Weber potential. Uniaxial tensile deformation along the zig-zag and armchair directions was considered for η = 1, 2, 3 or 4. The Young's modulus and fracture strength were higher along the armchair direction than the zig-zag direction, due to the corrugated structure along the zig-zag axis. The fracture resistance was very sensitive to temperature, while its dependence upon strain-rate was relatively small.

Molecular dynamics simulation was used[313] to study the effect of vacancies and impurities upon the tensile properties of monolayer material under uniaxial tension. The Young's modulus of nanosheet at 1K was 165.03N/m in the zig-zag direction and 390.02N/m in the armchair direction. The Poisson ratio was negative in the direction perpendicular to the atomic plane. The strength and ultimate strain decreased with increasing temperature. Stress-concentration around vacancy defects led to initial crack formation at the vacancy and to a reduction in fracture strength. Upon increasing the dimensions of vacancies projected onto the direction perpendicular to the tensile direction, the fracture strength, fracture strain and Young's modulus decreased. The strength was reduced by substitutional carbon atoms. When the defect-angle was 60°, the fracture of nanosheet with carbon atoms occurred more easily than it did when the angle was 0°, for both the zig-zag and armchair tensile directions.

A first-principles study was made[314] of the mechanical properties of two borophene structures, B1 and B2, when subjected to 3 three types of strain. The properties of B1 and B2 were highly anisotropic. The Young's moduli were 354 and 145N/m for B1 and 147 and 123N/m for B2, along 2 different directions.

Reactive molecular dynamics simulations were used[315] to investigate the fracture of nanofilms. Mode-I fracture of polycrystalline material with different grain sizes and crack propagation in sheets with various initial edge-crack sizes under mixed-mode loading were studied. A single primary crack bifurcated into two or more branches under mixed-mode loading conditions when the speed of the crack exceeded a certain level. This tendency was independent of the initial crack size and the strain-rate.

Four atomistic configurations of monolayers have been analyzed[316]: an ideal perfect sheet made only from regular triangular motifs, a sheet made from the ideal structure of the B_{36} molecule, a sheet containing uniformly distributed regular hexagonal holes and a sheet with hexagonal holes distributed in lines. Each sheet comprised an almost square shape of the same size. On the basis of computed tensile stress-strain curves, mechanical properties along the zig-zag and armchair directions were estimated. The Young's modulus, Poisson ratio, tensile strength and failure strain were related to the hexagonal hole density.

First-principles density functional theory calculations were used to investigate the properties of 5 types of single-layer sheets[317]. An analysis was made of the effect of the loading direction and vacancy content upon the mechanical response. Depending upon the atomic configuration and the loading direction, the modulus could range from 163 to 382GPa. The ultimate tensile strength ranged from 13.5 to about 22.8GPa, with the corresponding strains ranging from 0.1 to 0.21.

Tensile, shear and nano-indentation failure processes were simulated by using molecular dynamics methods[318]. The mechanical response and deformation mechanism under various loads were analyzed in terms of the change in B-B bond-length with strain/indentation depth. The tensile properties were notably anisotropic, in that the Young's modulus and strength along the armchair direction were much higher than those along the zig-zag direction. The anisotropy of the shear properties was negligible. The results were explained in terms of the differing contributions made by strong B-B σ-bonds and weak multi-centre bonds when stretched in different directions. The material also exhibited differing mechanical responses under spherical and cylindrical indentation: the force at failure under spherical indentation was much smaller than the value under cylindrical indentation. The intrinsic mechanical parameters under spherical indentation could not be accurately determined, because of the anisotropic characteristics. Under cylindrical indentation, the material exhibited similar anisotropic characteristics to those under tension. Parameters such as the Young's modulus were consistent with those for tension. The Young's modulus decreased slightly with increasing temperature, while the strain-rate had essentially no effect upon the Young's modulus.

The properties of material comprising 16% of hollow hexagons were studied by means of first-principles calculations and were predicted[319] to have in-plane modulus of up to 210N/m and a bending stiffness as low as 0.39eV. The Foppl–von Karman number per unit area attained $568/nm^2$, making borophene one of the most flexible materials. The specific modulus of $346m^2/s^2$ and ideal strength of $16N/m$ approached those of graphene. The delocalized multi-centre chemical bonding favoured structural phase transitions under tension and resulted in very small breaking strains but a highly ductile fracture behaviour. These properties could be modified by varying the hollow-hexagon concentration.

Density functional theory and semi-empirical van der Waals dispersion corrections were used[320] to investigate the stability of boron vacancies, substitutional and interstitial atoms (hydrogen, carbon, boron, nitrogen, oxygen) and $\Sigma5$ tilt grain boundaries. Pristine material exhibited a marked anisotropy of the in-plane Young's modulus and Poisson ratio, due to strong and highly coordinated B-B bonds. The concentration of boron vacancies and $\Sigma5$ grain boundaries could be high, because their formation energies were as low as 0.10 and 0.06eV/Å, respectively. The mechanical strength was markedly reduced by the defects. The anisotropy of the Poisson ratio could be affected by some of them. That is, adsorbed hydrogen and substitutional carbon could produce a negative Poisson ratio, while substitutional carbon and nitrogen could greatly increase the Poisson ratio.

The ideal tensile strength of double-layered borophene with lithium-storage, under uniaxial tension, was predicted[321] by using first-principles density functional theory calculations. The effect of the layer number upon the elastic modulus was negligible. Double-layered material was more stable than single-layer material. When the number of lithium atom storage in double-layered material was 12, the moduli in the X and Y directions were 768 and 241GPa, respectively.

The effect of halogen adsorption upon the mechanical properties of β_{12} borophene sheet was determined[322] by using density functional theory. The adsorption of fluorine, chlorine, bromine and iodine led to a decrease in the Young's modulus.

The uniform coating of borophene with fluorine can lead to derivatives having novel properties. First-principles molecular dynamics simulations indicated that, although fully fluorinated material, BF, could not be obtained, stable configurations, B_4F and B_2F, having lower coverages were possible[323]. Fluorination markedly altered the mechanical properties, with direction-dependent variations in the Young's modulus and a change in Poisson ratio from negative to positive.

The hexagonal lattice of borophene can be connected with other hexagonal lattices by using hydrogen atoms as connecting bridges, and hydrogenation greatly improved the Young's modulus of armchair material[324]. An increase in temperature decreased the Young's modulus. The presence of a vacancy also greatly decreased the Young's modulus; a reduction which was less marked for hydrogenated material.

Materials Research Forum LLC

https://doi.org/10.21741/9781644903650

The mechanical properties and phonon dispersion of fully hydrogenated borophene, i.e. borophane, under strain were studied by using first-principles calculations[325]. Uniaxial tensile straining along the a- and b-directions, and biaxial tensile straining, was considered. The mechanical properties were very anisotropic, in that the ultimate tensile strain along the a-direction was only 0.12, but could be up to 0.30 along the b-direction. When compared with borophene and other 2-dimensional materials, borophane thus exhibited the most marked anisotropy of in-plane ultimate strain. The phonon dispersions for the 3 applied strains indicated that borophane could suffer up to 5% and 15% uniaxial tensile strain along the a- and b-directions, respectively, and 9% biaxial tensile strain. This suggested that mechanical failure was likely to be due to phonon instability.

Plumbene

This 2-dimensional allotrope of lead is again analogous to graphene, silicene, germanene and stanene with regard to its single-layered and hexagonal atomic structure. Molecular dynamics simulations were used[326] to study the effects of sample-size, temperature and strain-rate upon the mechanical properties. The properties of single-layer sheet were several times better than those of bulk lead, thus making it a potential reinforcement for high-strength nano-composites. The yield strength of the sheet decreased sharply for sample-lengths of 700 to 900Å, and then increased for lengths of up to 1500Å; with a marked increase at 1100Å. The latter was attributed to the high elasticity at that sheet-size. The ultimate tensile strength increased sharply from 700 to 900Å, remained constant from 900 to 1100Å and decreased gradually between 1100 and 1500Å, with a sharp drop between 1300 and 1500Å. The Young's modulus meanwhile decreased sharply from 700 to 900Å, increased linearly between 900 and 1300Å and then suddenly decreased between 1300 and 1500Å. A sheet-size of 900Å offered the lowest resistance to elastic deformation. The yield strength gradually increased from 318 to 358K, with a sharp increase from 338 to 358K. It decreased from 358 to 398K, with a sharp decrease from 378 to 398K. The sheet-size of 1100Å exhibited the highest yield strength at 358K, but this was lower than the yield strength at room temperature. This was attributed to the increase in atomic vibration with increasing temperature. The ultimate tensile strength decreased from 318 to 338K, remained constant from 338 to 378K and decreased from 378 to 398K. The ultimate tensile strength was not affected by temperatures of 338 to 378K. The Young's modulus meanwhile gradually increased from 318 to 398K, with a sharp increase from 338 to 358K and from 378 to 398K. The mechanical properties were also investigated[327] by means of tensile modeling. The Young's modulus was estimated by using bending, oscillation and equilibrium molecular dynamics methods. The bulk modulus and Poisson ratio were also estimated by using equilibrium molecular dynamic methods. The bulk modulus, Poisson ratio and Young's modulus were computed to be 4.2994MPa, 0.33 and 4.1887MPa, respectively. Depending upon the assumed testing method, E could range from 0.6516MPa to 110.07MPa. The Poisson ratio of 0.33 was close to that (0.3) of silicene. The effects of increasing temperature and sample size were also studied via molecular dynamics simulations of plumbene sheet[328]. The Young's modulus was estimated via the bending and oscillation modeling of plumbene under various loading conditions. First-principles methods were used to estimate the mechanical properties of monolayers under external strains and electric fields and in the presence of Stone-Wales defects[329]. A plumbene monolayer had the lowest Young's modulus and ideal strength of the group-IV mono-elemental monolayers. Tensile straining softened optical phonon modes to the extent that the gap between acoustic and phonon modes disappeared at high strains, and the failure mechanism in plumbene monolayers was phonon instability. Biaxial tensile straining could not open a band-gap, but a perpendicular electric field could open a small band-gap at the K-point. A monolayer could recover from Stone-Wales defects, but there was a transition from a low-buckle monolayer to a high-buckle one.

About the Author

Dr. Fisher has wide knowledge and experience of the fields of engineering, metallurgy and solid-state physics, beginning with work at Rolls-Royce Aero Engines on turbine-blade research, related to the Concord supersonic passenger-aircraft project, which led to a BSc degree (1971) from the University of Wales. This was followed by theoretical and experimental work on the directional solidification of eutectic alloys having the ultimate aim of developing composite turbine blades. This work led to a doctoral degree (1978) from the Swiss Federal Institute of Technology (Lausanne). He then acted for many years as an editor of various academic journals, in particular *Defect and Diffusion Forum*. In recent years he has specialized in writing monographs which introduce readers to the most rapidly developing ideas in the fields of engineering, metallurgy and solid-state physics. He is co-author of the widely-cited student textbook, *Fundamentals of Solidification*. Google Scholar credits him with 8687 citations and a lifetime h-index of 14.

References

[1] Storch I.R., De Alba R., Adiga V.P., Abhilash T.S., Barton R.A., Craighead H.G., Parpia J.M., McEuen P.L., Physical Review B, 98[8] 2018, 085408. https://doi.org/10.1103/PhysRevB.98.085408

[2] Han J., Ryu S., Kim D.K., Woo W., Sohn D., EPL, 114[6] 2016, 68001. https://doi.org/10.1209/0295-5075/114/68001

[3] Li H., Gürbüz E., Haldar S., Hussain T., Zheng X., Ye X., Makumi S.W., Duan T., Jafri S.H.M., Daukiya L., Simon L., Karton A., Sanyal B., Leifer K., Applied Physics Letters, 123[5] 2023, 053102. https://doi.org/10.1063/5.0157104

[4] Braun M., Arca F., Ariza M.P., International Journal of Mechanical Sciences, 209, 2021, 106702. https://doi.org/10.1016/j.ijmecsci.2021.106702

[5] Chen X., Yi C., Ke C., Applied Physics Letters, 106[10] 2015, 101907. https://doi.org/10.1063/1.4915075

[6] Garg A., Vijayaraghavan V., Wong C.H., Tai K., Sumithra K., Gao L., Singru P.M., Simulation Modelling Practice and Theory, 48, 2014, 93-111. https://doi.org/10.1016/j.simpat.2014.07.008

[7] Calis M., Boddeti N., Bunch J.S., Nanoscale, 17[4] 2024, 2235-2242. https://doi.org/10.1039/D4NR04214J

[8] Liu X., Metcalf T.H., Robinson J.T., Perkins F.K., Houston B.H., Solid State Phenomena, 184, 2012, 319-324. https://doi.org/10.4028/www.scientific.net/SSP.184.319

[9] Shi J.X., Natsuki T., Lei X.W., Ni Q.Q., Applied Physics Letters, 104[22] 2014, 223101. https://doi.org/10.1063/1.4880729

[10] Thomas S., Mrudul M.S., Ajith K.M., Valsakumar M.C., Journal of Physics - Conference Series, 759[1] 2016, 012048. https://doi.org/10.1088/1742-6596/759/1/012048

[11] Jing N., Xue Q., Ling C., Shan M., Zhang T., Zhou X., Jiao Z., RSC Advances, 2[24] 2012, 9124-9129. https://doi.org/10.1039/c2ra21228e

[12] Zeinalipour-Yazdi C.D., Christofides C., Journal of Applied Physics, 106[5] 2009, 054318. https://doi.org/10.1063/1.3211944

[13] Zhu J., He M., Qiu F., Chinese Journal of Chemistry, 30[7] 2012, 1399-1404. https://doi.org/10.1002/cjoc.201200505

[14] Lee J.U., Yoon D., Cheong H., Nano Letters, 12[9] 2012, 4444-4448. https://doi.org/10.1021/nl301073q

[15] Zhong T., Li J., Zhang K., Journal of Applied Physics, 125[17], 2019, 175110. https://doi.org/10.1063/1.5091753

[16] Tan X., Wu J., Zhang K., Peng X., Sun L., Zhong J., Applied Physics Letters, 102[7[
2013, 071908. https://doi.org/10.1063/1.4793191

[17] Sajadi B., Wahls S., Hemert S.V., Belardinelli P., Steeneken P.G., Alijani F., Journal
of the Mechanics and Physics of Solids, 122, 2019, 161-176.
https://doi.org/10.1016/j.jmps.2018.09.013

[18] Chu L., Shi J., Braun R., Physica E, 110, 2019, 115-122.
https://doi.org/10.1016/j.physe.2019.01.023

[19] Dettori R., Cadelano E., Colombo L., Journal of Physics - Condensed Matter, 24[10]
2012, 104020. https://doi.org/10.1088/0953-8984/24/10/104020

[20] Shodja H.M., Delfani M.R., Acta Mechanica, 222[1-2] 2011, 91-101.
https://doi.org/10.1007/s00707-011-0528-5

[21] Xiang L., Ma S.Y., Wang F., Zhang K., Journal of Physics D, 48[39] 2015, 395305.
https://doi.org/10.1088/0022-3727/48/39/395305

[22] Jiang J.W., Wang J.S., Li B., Physical Review B, 80[11] 2009, 113405.
https://doi.org/10.1103/PhysRevB.80.113405

[23] Liang Y., Han Q., Huan S., Journal of Thermal Stresses, 38[8] 2015, 926-933.
https://doi.org/10.1080/01495739.2015.1040317

[24] Damasceno D.A., Rajapakse R.K.N.D., Mesquita E., Pavanello R., Acta Mechanica,
231[8] 2020, 3387-3404. https://doi.org/10.1007/s00707-020-02715-6

[25] Mazilova T.I., Sadanov E.V., Mikhailovskij I.M., Materials Letters, 242, 2019, 17-
19. https://doi.org/10.1016/j.matlet.2019.01.051

[26] Kochnev A.S., Ovidko I.A., Semenov B.N., Reviews on Advanced Materials
Science, 37[1-2] 2014, 105-110.

[27] Qin X., Yan W., Guo X., Gao T., RSC Advances, 8[31] 2018, 17034-17043.
https://doi.org/10.1039/C8RA02415D

[28] Debroy S., Acharyya S.G., Acharyya A., Transactions of the Indian Institute of
Metals, 74[7] 2021, 1729-1739. https://doi.org/10.1007/s12666-021-02258-x

[29] Han J., Pugno N.M., Ryu S., Nanoscale, 7[38] 2015, 15672-15679.
https://doi.org/10.1039/C5NR04134A

[30] Giannopoulos G.I., Avntoulla G.S., Proceedings of the Institution of Mechanical
Engineers N, 231[2] 2017, 67-73. https://doi.org/10.1177/2397791417712845

[31] Kashani H., Ito Y., Han J., Liu P., Chen M., Science Advances, 5[2] 2019, eaat6951.
https://doi.org/10.1126/sciadv.aat6951

[32] Sun H., Mukherjee S., Shi Z., Singh C.V., Carbon, 143, 2019, 752-761.
https://doi.org/10.1016/j.carbon.2018.11.079

[33] Jaddi S., Malik M.W., Wang B., Pugno N.M., Zeng Y., Coulombier M., Raskin J.P., Pardoen T., Nature Communications, 15[1] 2024, 5863. https://doi.org/10.1038/s41467-024-49426-3

[34] Sutrakar V.K., Javvaji B., Budarapu P.R., Applied Physics A, 127[12] 2021, 949. https://doi.org/10.1007/s00339-021-05047-x

[35] Arshad M.U., Gan Y., Wei C., Li J., Wu C., Naraghi M., Small, 19[39] 2023, 2302499. https://doi.org/10.1002/smll.202302499

[36] Brandão W.H.S., Aguiar A.L., De Sousa J.M., Chemical Physics Letters, 778, 2021, 138793. https://doi.org/10.1016/j.cplett.2021.138793

[37] Yeh Y.K., Hwu C., Engineering Fracture Mechanics, 176, 2017, 326-342. https://doi.org/10.1016/j.engfracmech.2017.03.039

[38] Hyong C.J., Cayll D., Ladner I., Gorman J., Cullinan M., Engineering Fracture Mechanics, 295, 2024, 109798. https://doi.org/10.1016/j.engfracmech.2023.109798

[39] Feng S., Cao K., Gao Y., Han Y., Liu Z., Lu Y., Xu Z., Communications Materials, 3[1] 2022, 28. https://doi.org/10.1038/s43246-022-00252-4

[40] Mai N.T., Choi S.T., International Journal of Solids and Structures, 138, 2018, 205-216. https://doi.org/10.1016/j.ijsolstr.2018.01.013

[41] Zhang Z., Zhang X., Wang Y., Wang Y., Zhang Y., Xu C., Zou Z., Wu Z., Xia Y., Zhao P., Wang H.T., ACS Nano, 13[9] 2019, 10327-10332. https://doi.org/10.1021/acsnano.9b03999

[42] Wood J.D., Harvey C.M., Wang S., Nature Communications, 8[1] 2017, 1952. https://doi.org/10.1038/s41467-017-02115-w

[43] Sheng L., Wei T., Liang Y., Jiang L., Qu L., Fan Z., Carbon, 120, 2017, 17-22. https://doi.org/10.1016/j.carbon.2017.05.033

[44] Tian Q., Xu Z., Liu Y., Fang B., Peng L., Xi J., Li Z., Gao C., Nanoscale, 9[34] 2017, 12335-12342. https://doi.org/10.1039/C7NR03895J

[45] Cui T., Mukherjee S., Sudeep P.M., Colas G., Najafi F., Tam J., Ajayan P.M., Singh C.V., Sun Y., Filleter T., Nature Materials, 19[4] 2020, 405-411. https://doi.org/10.1038/s41563-019-0586-y

[46] Yang Z., Huang Y., Bao H., Xu K., Ma F., Journal of Materials Science, 52[18] 2017, 10871-10878. https://doi.org/10.1007/s10853-017-1269-1

[47] Cui T., Yip K., Hassan A., Wang G., Liu X., Sun Y., Filleter T., Science Advances, 6[42] 2020, eabb1335. https://doi.org/10.1126/sciadv.abb1335

[48] Kara A., Léandri C., Dávila M.E., De Padova P., Ealet B., Oughaddou H., Aufray B., Le Lay G., Journal of Superconductivity and Novel Magnetism, 22[3] 2009, 259-263. https://doi.org/10.1007/s10948-008-0427-8

[49] O'Hare A., Kusmartsev F.V., Kugel K.I., Nano Letters, 12[2] 2012, 1045-1052. https://doi.org/10.1021/nl204283q

[50] Song Y.L., Zhang Y., Zhang J.M., Lu D.B., Applied Surface Science, 256[21] 2010, 6313-6317. https://doi.org/10.1016/j.apsusc.2010.04.009

[51] Sun M., Ren Q., Wang S., Yu, J., Tang, W., Journal of Physics D, 49[44] 2016, 445305. https://doi.org/10.1088/0022-3727/49/44/445305

[52] Fang T.H., Shen C.Y., Fan Y.C., Chang W.J., Superlattices and Microstructures, 129, 2019, 124-129. https://doi.org/10.1016/j.spmi.2019.03.021

[53] Qu Z., Xu M., Lin S., Liang Y., Yuan X., Wang F., Hao J., Li Y., Nanoscale, 14[29] 2022, 10573-10580. https://doi.org/10.1039/D2NR01465C

[54] Roman R.E., Cranford S.W., Mechanics of Materials, 133, 2019, 13-25. https://doi.org/10.1016/j.mechmat.2019.01.014

[55] Yoo S., Lee B., Kang K., Nanotechnology, 32[29] 2021, 295702. https://doi.org/10.1088/1361-6528/abf26d

[56] Vali M., Safa S., Dideban D., Journal of Materials Science: Materials in Electronics, 29[23] 2018, 20522-20529. https://doi.org/10.1007/s10854-018-0189-4

[57] Nickabadi S., Ansari R., Rouhi S., Aghdasi P., Journal of Molecular Modeling, 27[6] 2021, 190. https://doi.org/10.1007/s00894-021-04780-7

[58] Nickabadi S., Ansari R., Rouhi S., Acta Mechanica, 231[10] 2020, 4351-4363. https://doi.org/10.1007/s00707-020-02757-w

[59] Das D.K., Sarkar J., Journal of Applied Physics, 123[4] 2018, 044304. https://doi.org/10.1063/1.5009084

[60] Chung J.Y., Sorkin V., Pei Q.X., Chiu C.H., Zhang Y.W., Journal of Physics D, 50[34] 2017, 345302. https://doi.org/10.1088/1361-6463/aa7938

[61] Rouhi S., Pourmirzaagha H., Farzin A., Materials Research Express, 6[8] 2019, 085004. https://doi.org/10.1088/2053-1591/ab1b81

[62] Peng Q., De S., Nanoscale, 6[20] 2014, 12071-12079. https://doi.org/10.1039/C4NR01831A

[63] Zhao H., Physics Letters A, 376[46] 2012, 3546-3550. https://doi.org/10.1016/j.physleta.2012.10.024

[64] Cheng G., Liu P.F., Li Z.T., Chinese Physics B, 22[4] 2013, 046201. https://doi.org/10.1088/1674-1056/22/4/046201

[65] Gargeya B.S.K., Pal S., Materials Today - Proceedings 18, 2019, 1401-1410. https://doi.org/10.1016/j.matpr.2019.06.606

[66] Le M.Q., Nguyen D.T., Applied Physics A, 118[4] 2025, 1437-1445. https://doi.org/10.1007/s00339-014-8904-3

[67] Liu N., Becton M., Zhang L., Tang K., Wang X., Nanoscale Advances, 1[8] 2019, 2891-2900. https://doi.org/10.1039/C9NA00312F

[68] Le M.Q., Nguyen H.T., Bui T.L., Mechanics of Advanced Materials and Structures, 29[26] 2022, 4993-5005. https://doi.org/10.1080/15376494.2021.1945172

[69] Le M.Q., Computational Materials Science, 201, 2022, 110854. https://doi.org/10.1016/j.commatsci.2021.110854

[70] Pei Q.X., Sha Z.D., Zhang Y.Y., Zhang Y.W., Journal of Applied Physics, 115[2] 2014, 023519. https://doi.org/10.1063/1.4861736

[71] Liu N., Hong J., Pidaparti R., Wang X., 2D Materials, 3[3] 2016, 035008. https://doi.org/10.1088/2053-1583/3/3/035008

[72] Rouhi S., Computational Materials Science, 131, 2017, 275-285. https://doi.org/10.1016/j.commatsci.2017.02.007

[73] Fan Y.C., Fang T.H., Chen T.H., Nanomaterials, 6[7] 2016, 120. https://doi.org/10.3390/nano6070120

[74] Ding N., Wang H., Chen X., Wu C.M.L., RSC Advances, 7[17] 2017, 10306-10315. https://doi.org/10.1039/C6RA27291F

[75] Davydov S.Y., Physics of the Solid State, 52[1] 2010, 184-187. https://doi.org/10.1134/S1063783410010312

[76] Roman R.E., Cranford S.W., Computational Materials Science, 82, 2014, 50-55. https://doi.org/10.1016/j.commatsci.2013.09.030

[77] Botari T., Perim E., Autreto P.A.S., Van Duin A.C.T., Paupitz R., Galvao D.S., Physical Chemistry Chemical Physics, 16[36] 2014, 19417-19423. https://doi.org/10.1039/C4CP02902J

[78] Yang C., Yu Z., Lu P., Liu Y., Ye H., Gao T., Computational Materials Science, 95, 2014, 420-428. https://doi.org/10.1016/j.commatsci.2014.07.046

[79] Peng Q., Wen X., De S., RSC Advances, 3[33] 2013, 13772-13781. https://doi.org/10.1039/c3ra41347k

[80] Qin R., Wang C.H., Zhu W., Zhang Y., AIP Advances, 2[2] 2012, 022159. https://doi.org/10.1063/1.4732134

[81] John R., Merlin B., Crystal Structure Theory and Applications, 5, 2016, 43-55. https://doi.org/10.4236/csta.2016.5304

[82] Ince A., Erkoç S., Physica Status Solidi B, 249[1] 2012, 74-81. https://doi.org/10.1002/pssb.201147267

[83] Ansari R., Rouhi S., Ajori S., Superlattices and Microstructures, 65, 2014, 64-70. https://doi.org/10.1016/j.spmi.2013.10.039

[84] Chávez-Castillo M.R., Rodríguez-Meza M.A., Meza-Montes L., RSC Advances, 5[116] 2015, 96052-96061. https://doi.org/10.1039/C5RA15312C

[85] Jing Y., Sun Y., Niu H., Shen J., Physica Status Solidi B, 250[8] 2013, 1505-1509. https://doi.org/10.1002/pssb.201349023

[86] Kaloni T.P., Cheng Y.C., Schwingenschlögl U., Journal of Applied Physics, 113[10] 2013, 104305. https://doi.org/10.1063/1.4794812

[87] Wang B., Wu J., Gu X., Yin H., Wei Y., Yang R., Dresselhaus M., Applied Physics Letters, 104[8] 2014, 081902. https://doi.org/10.1063/1.4866415

[88] Huang L.F., Gong P.L., Zeng Z., Physical Review B, 91[20] 2015, 205433. https://doi.org/10.1103/PhysRevB.91.205433

[89] Yuan X., Lin G., Wang Y., Molecular Simulation, 42[14] 2016, 1157-1164. https://doi.org/10.1080/08927022.2016.1148266

[90] Lee C.C., Fleurence A., Friedlein R., Yamada-Takamura Y., Ozaki T., Physical Review B, 88[16] 2013, 165404. https://doi.org/10.1103/PhysRevB.88.165404

[91] Fleurence A., Yamada-Takamura Y., Applied Physics Letters, 110[4] 2017, 041601. https://doi.org/10.1063/1.4974467

[92] Li X.L., Liu G., Du T.Y., Zhao J., Wu M.S., Ouyang C.Y., Xu B., Acta Physica Sinica, 63[21] 2014, 217101. https://doi.org/10.7498/aps.63.217101

[93] Wang Y., Ding Y., Physica Status Solidi R, 7[6] 2013, 410-413. https://doi.org/10.1002/pssr.201307110

[94] Zhang W.B., Song Z.B., Dou L.M., Journal of Materials Chemistry C, 3[13] 2015, 3087. https://doi.org/10.1039/C4TC02758B

[95] Davydov S.Y., Physics of the Solid State, 54[3] 2012, 652-657. https://doi.org/10.1134/S1063783412030080

[96] Xu P., Yu Z., Yang C., Lu P., Liu Y., Ye H., Gao T., Superlattices and Microstructures, 75, 2014, 647-656. https://doi.org/10.1016/j.spmi.2014.08.022

[97] Barati M., Vazifehshenas T., Salavati-fard T., Farmanbar M., Computational Materials Science, 173, 2020, 109407. https://doi.org/10.1016/j.commatsci.2019.109407

[98] Hu M., Zhang X., Poulikakos D., Physical Review B, 87[19] 2013, 195417. https://doi.org/10.1103/PhysRevB.87.195417

[99] Gupta N., Verma R., Superlattices and Microstructures, 156, 2021, 106944. https://doi.org/10.1016/j.spmi.2021.106944

[100] Mohan B., Kumar A., Ahluwalia P.K., Physica E, 61, 2014, 40-47. https://doi.org/10.1016/j.physe.2014.03.013

[101] Pei Q.X., Zhang Y.W., Sha Z.D., Shenoy V.B., Journal of Applied Physics, 114[3] 2013, 033526. https://doi.org/10.1063/1.4815960

[102] Xie H., Ouyang T., Germaneau É., Qin G., Hu M., Bao H., Physical Review B, 93[7] 2016, 075404. https://doi.org/10.1103/PhysRevB.93.075404

[103] Khalkhali M., Rajabpour A., Khoeini F., Scientific Reports, 9[1] 2019, 5684. https://doi.org/10.1038/s41598-019-42187-w

[104] Ding B., Li X., Zhou W., Zhang G., Gao H., National Science Review, 8[9] 2021, nwaa220. https://doi.org/10.1093/nsr/nwaa220

[105] Chen G., Hu B., Wang Z., Tang D., International Journal of Thermophysics, 44[4] 2023, 60. https://doi.org/10.1007/s10765-023-03170-w

[106] Khalkhali M., Khoeini F., Rajabpour A., International Journal of Heat and Mass Transfer, 134, 2019, 503-510. https://doi.org/10.1016/j.ijheatmasstransfer.2019.01.074

[107] Liu H., Qin G., Hu M., Frontiers in Materials, 8, 2021, 683503. https://doi.org/10.3389/fmats.2021.683503

[108] Kuang Y.D., Lindsay L., Shi S.Q., Zheng G.P., Nanoscale, 8[6] 2016, 3760-3767. https://doi.org/10.1039/C5NR08231E

[109] Yan J.A., Gao S.P., Stein R., Coard G., Physical Review B, 91[24] 2015, 245403. https://doi.org/10.1103/PhysRevB.91.245403

[110 Wani A.F., Khandy S.A., Verma A.S., Dhiman S., Kaur K., Scientific Reports, 14[1] 2024, 29555. https://doi.org/10.1038/s41598-024-81138-y

[111] Sahin, H., Sivek, J., Li, S., Partoens, B., Peeters, F.M., Physical Review B, 88[4] 2013, 045434. https://doi.org/10.1103/PhysRevB.88.045434

[112] Ting, X., Rui, W., Shaofeng, W., Xiaozhi, W., Applied Surface Science, 383, 2016, 310-316. https://doi.org/10.1016/j.apsusc.2016.04.172

[113] Manjanath, A., Singh, A.K., Chemical Physics Letters, 592, 2014, 52-55. https://doi.org/10.1016/j.cplett.2013.12.010

[114] Zha, D., Chen, C., Wu, J., Wang, M., International Journal of Modern Physics B, 29[9] 2015, 1550061. https://doi.org/10.1142/S0217979215500617

[115] Iordanidou, K., Houssa, M., Van Den Broek, B., Pourtois, G., Afanasev, V.V., Stesmans, A., Journal of Physics - Condensed Matter, 28[3] 2016, 035302. https://doi.org/10.1088/0953-8984/28/3/035302

[116] Gao, J., Zhang, J., Liu, H., Zhang, Q., Zhao, J., Nanoscale, 5[20] 2013, 9785-9792. https://doi.org/10.1039/c3nr02826g

[117] Le, M.Q., Nguyen, D.T., Applied Physics A, 118[4] 2014, 1437-1445. https://doi.org/10.1007/s00339-014-8904-3

[118] Dong, H., Fang, D., Gong, B., Zhang, Y., Zhang, E., Zhang, S., Journal of Applied Physics, 117[6] 2015, 064307. https://doi.org/10.1063/1.4907582

[119] Balendhran, S., Walia, S., Nili, H., Sriram, S., Bhaskaran, M., Small, 11[6] 2015, 640-652. https://doi.org/10.1002/smll.201402041

[120] Dimoulas, A., Microelectronic Engineering, 131, 2015, 68-78. https://doi.org/10.1016/j.mee.2014.08.013

[121] Persichetti, L., Jardali, F., Vach, H., Sgarlata, A., Berbezier, I., Crescenzi, M.D., Balzarotti, A., Journal of Physical Chemistry Letters, 7[16] 2016, 3246-3251. https://doi.org/10.1021/acs.jpclett.6b01284

[122] Zhang, L., Bampoulis, P., Rudenko, A.N., Yao, Q., Van Houselt, A., Poelsema, B., Katsnelson, M.I., Zandvliet, H.J.W., Physical Review Letters, 116[25] 2016, 256804 https://doi.org/10.1103/PhysRevLett.116.256804

[123] Gou, J., Zhong, Q., Sheng, S., Li, W., Cheng, P., Li, H., Chen, L., Wu, K., 2D Materials, 3[4] 2016, 045005. https://doi.org/10.1088/2053-1583/3/4/045005

[124] Dávila, M.E., Le Lay, G., Scientific Reports, 6, 2016, 20714. https://doi.org/10.1038/srep20714

[125] Dávila, M.E., Xian, L., Cahangirov, S., Rubio, A., Le Lay, G., New Journal of Physics, 16, 2014, 095002. https://doi.org/10.1088/1367-2630/16/9/095002

[126] Naji, S., Belhaj, A., Labrim, H., Benyoussef, A., El Kenz, A., Modern Physics Letters B, 27[29] 2013, 1350212. https://doi.org/10.1142/S0217984913502126

[127] Li, L., Lu, S.Z., Pan, J., Qin, Z., Wang, Y.Q., Wang, Y., Cao, G.Y., Du, S., Gao, H.J., Advanced Materials, 26[28] 2014, 4820-4824. https://doi.org/10.1002/adma.201400909

[128] Özçelik, V.O., Durgun, E., Ciraci, S., Journal of Physical Chemistry Letters, 5[15] 2014, 2694-2699. https://doi.org/10.1021/jz500977v

[129] Roome, N.J., Carey, J.D., ACS Applied Materials and Interfaces, 6[10] 2014, 7743-7750. https://doi.org/10.1021/am501022x

[130] Cai Y., Chuu C.P., Wei C.M., Chou M.Y., Physical Review B, 88[24] 2013, 245408. https://doi.org/10.1103/PhysRevB.88.245408

[131] Tsai H.S., Chen Y.Z., Medina H., Su T.Y., Chou T.S., Chen Y.H., Chueh Y.L., Liang J.H., Physical Chemistry Chemical Physics, 17[33] 2015, 21389-21393. https://doi.org/10.1039/C5CP02469B

[132] Li P., Cao J., Guo Z.X., Journal of Materials Chemistry C, 4[8] 2016, 1736-1740. https://doi.org/10.1039/C5TC03442F

[133] Wang Y., Ding Y., Solid State Communications, 155, 2013, 6-11. https://doi.org/10.1016/j.ssc.2012.10.044

[134] Tomberli, B., Rahemtulla, A., Kim, E., Roorda, S., Kycia, S., Physical Review B, 92[6] 2015, 064204. https://doi.org/10.1103/PhysRevB.92.064204

[135] Nijamudheen, A., Bhattacharjee, R., Choudhury, S., Datta, A., Journal of Physical Chemistry C, 119[7] 2015, 3802-3809. https://doi.org/10.1021/jp511488m

[136] Trivedi, S., Srivastava, A., Kurchania, R., Journal of Computational and Theoretical Nanoscience, 11[3] 2014, 781-788. https://doi.org/10.1166/jctn.2014.3428

[137] Acun, A., Zhang, L., Bampoulis, P., Farmanbar, M., Van Houselt, A., Rudenko, A.N., Lingenfelder, M., Brocks, G., Poelsema, B., Katsnelson, M.I., Zandvliet, H.J.W., Journal of Physics - Condensed Matter, 27[44] 2015, 443002. https://doi.org/10.1088/0953-8984/27/44/443002

[138] Stephan, R., Hanf, M.C., Derivaz, M., Dentel, D., Asensio, M.C., Avila, J., Mehdaoui, A., Sonnet, P., Pirri, C., Journal of Physical Chemistry C, 120[3] 2016, 1580-1585. https://doi.org/10.1021/acs.jpcc.5b10307

[139] D'Acapito, F., Torrengo, S., Xenogiannopoulou, E., Tsipas, P., Marquez Velasco, J., Tsoutsou, D., Dimoulas, A., Journal of Physics - Condensed Matter, 28[4] 2016, 045002. https://doi.org/10.1088/0953-8984/28/4/045002

[140] Liang, P., Liu, Y., Xing, S., Shu, H., Tai, B., Solid State Communications, 226, 2016, 19-24. https://doi.org/10.1016/j.ssc.2015.10.021

[141] Yuan, J., Tang, C., Zhong, J., Mao, Y., Applied Surface Science, 360, 2016, 707-714. https://doi.org/10.1016/j.apsusc.2015.11.050

[142] Kaloni, T.P., Schwingenschlögl, U., Chemical Physics Letters, 583, 2013, 137-140. https://doi.org/10.1016/j.cplett.2013.08.001

[143] Huang, L.F., Gong, P.L., Zeng, Z., Physical Review B, 91[20] 2015, 205433. https://doi.org/10.1103/PhysRevB.91.205433

[144] Singh N., Schwingenschlögl U., Physic. https://doi.org/10.1002/pssr.201409025a Status Solidi R, 8[4] 2014, 353-356.

[145] Lin C.H., Huang A., Pai W.W., Chen W.C., Chen T.Y., Chang T.R., Yukawa R., Cheng C.M., Mou C.Y., Matsuda I., Chiang T.C., Jeng H.T., Tang S.J., Physical Review Materials, 2[2] 2018, 024003. https://doi.org/10.1103/PhysRevMaterials.2.024003

[146] Shiraz A.K., Goharrizi A.Y., Hamidi S.M., Physica E, 107, 2019, 150-153. https://doi.org/10.1016/j.physe.2018.11.019

[147] Mortazavi, B., Rahaman, O., Makaremi, M., Dianat, A., Cuniberti, G., Rabczuk, T., Physica E, 87, 2017, 228-232. https://doi.org/10.1016/j.physe.2016.10.047

[148] John, R., Merlin, B., Crystal Structure Theory and Applications, 5, 2016, 43-55. https://doi.org/10.4236/csta.2016.5304

[149] Kaloni T.P., Schwingenschlögl U., Chemical Physics Letters, 583, 2013, 137-140. https://doi.org/10.1016/j.cplett.2013.08.001

[150] Mortazavi B., Rahaman O., Makaremi M., Dianat A., Cuniberti G., Rabczuk T., Physica E, 87, 2017, 228-232. https://doi.org/10.1016/j.physe.2016.10.047

[151] Kaloni, T.P., Journal of Physical Chemistry C, 118[43] 2014, 25200-25208. https://doi.org/10.1021/jp5058644

[152] Pang, Q., Li, L., Zhang, C.L., Wei, X.M., Song, Y.L., Materials Chemistry and Physics, 160, 2015, 96-104. https://doi.org/10.1016/j.matchemphys.2015.04.011

[153] Pang, Q., Zhang, C.L., Li, L., Fu, Z.Q., Wei, X.M., Song, Y.L., Applied Surface Science, 314, 2014, 15-20. https://doi.org/10.1016/j.apsusc.2014.06.138

[154] Pang, Q., Li, L., Zhang, L.L., Zhang, C.L., Song, Y.L., Canadian Journal of Physics, 93[11] 2015, 1310-1318. https://doi.org/10.1139/cjp-2015-0206

[155] Li, S.S., Zhang, C.W., Ji, W.X., Li, F., Wang, P.J., Hu, S.J., Yan, S.S., Liu, Y.S., Physical Chemistry Chemical Physics, 16[30] 2014, 15968-15978. https://doi.org/10.1039/C4CP01211A

[156] Van Den Broek, B., Houssa, M., Scalise, E., Pourtois, G., Afanasev, V.V., Stesmans, A., Applied Surface Science, 291, 2014, 104-108. https://doi.org/10.1016/j.apsusc.2013.09.032

[157] Gürel, H.H., Özçelik, V.O., Ciraci, S., Journal of Physics - Condensed Matter, 25[30] 2013, 305007. https://doi.org/10.1088/0953-8984/25/30/305007

[158] Ongun Özçelik, V., Kecik, D., Durgun, E., Ciraci, S., Journal of Physical Chemistry C, 119[1] 2015, 845-853. https://doi.org/10.1021/jp5106554

[159] Rubio-Pereda, P., Takeuchi, N., Journal of Physical Chemistry C, 119[50] 2015, 27995-28004. https://doi.org/10.1021/acs.jpcc.5b08370

[160] Houssa, M., Scalise, E., Sankaran, K., Pourtois, G., Afanasev, V.V., Stesmans, A., Applied Physics Letters, 98[22] 2011, 223107. https://doi.org/10.1063/1.3595682

[161] Wella, S.A., Suprijadi, ARPN Journal of Engineering and Applied Sciences, 10[18] 2015, 8250-8253

[162] Jamdagni, P., Kumar, A., Sharma, M., Thakur, A., Ahluwalia, P.K., Physica E, 85, 2017, 65-73. https://doi.org/10.1016/j.physe.2016.08.015

[163] Xiao, P., Fan, X.L., Liu, L.M., Computational Materials Science, 92, 2014, 244-252. https://doi.org/10.1016/j.commatsci.2014.05.041

[164[Wang, X.Q., Li, H.D., Wang, J.T., Physical Chemistry Chemical Physics, 14[9] 2012, 3031-3036. https://doi.org/10.1039/c2cp23385a

[165] Rupp, C.J., Chakraborty, S., Ahuja, R., Baierle, R.J., Physical Chemistry Chemical Physics, 17[34] 2015, 22210-22216. https://doi.org/10.1039/C5CP03489B

[166] Liu, G., Liu, S.B., Xu, B., Ouyang, C.Y., Song, H.Y., Journal of Applied Physics, 118[12] 2015, 124303. https://doi.org/10.1063/1.4931057

[167] Xia, W., Hu, W., Li, Z., Yang, J., Physical Chemistry Chemical Physics, 16[41] 2014, 22495-22498. https://doi.org/10.1039/C4CP03292F

[168] Pang, Q., Li, L., Gao, D.L., Chai, R.P., Zhang, C.L., Song, Y.L., Physica E, 88, 2017, 237-242. https://doi.org/10.1016/j.physe.2017.01.018

[169] García, G., Atilhan, M., Aparicio, S., Physical Chemistry Chemical Physics, 17[25] 2015, 16315-16326. https://doi.org/10.1039/C5CP02432C

[170] Padilha, J.E., Pontes, R.B., Solid State Communications, 225, 2016, 38-43. https://doi.org/10.1016/j.ssc.2015.10.019

[171] Scalise, E., Houssa, M., Pourtois, G., van den Broek, B., Afanasev, V., Stesmans, A., Nano Research, 6[1] 2013, 19-28. https://doi.org/10.1007/s12274-012-0277-3

[172] Saxena, S., Chaudhary, R.P., Shukla, S., Scientific Reports, 6, 2016, 31073. https://doi.org/10.1038/srep31073

[173] Zhu, F.F., Chen, W.J., Xu, Y., Gao, C.L., Guan, D.D., Liu, C.H., Qian, D., Zhang, S.C., Jia, J.F., Nature Materials, 14[10] 2015, 1020-1025. https://doi.org/10.1038/nmat4384

[174] van den Broek, B., Houssa, M., Pourtois, G., Afanasev, V.V., Stesmans, A., Physica Status Solidi R, 8[11] 2014, 931-934. https://doi.org/10.1002/pssr.201400073

[175] Gao, J., Zhang, G., Zhang, Y.W., Scientific Reports, 6, 2016, 29107. https://doi.org/10.1038/srep24609

[176] Matusalem, F., Marques, M., Teles, L.K., Bechstedt, F., Physical Review B, 92[4] 2015, 045436. https://doi.org/10.1103/PhysRevB.92.045436

[177] Mogulkoc, A., Modarresi, M., Kandemir, B.S., Roknabadi, M.R., Physica Status Solidi B, 253[2] 2016, 300-307. https://doi.org/10.1002/pssb.201552341

[178] Van Den Broek, B., Houssa, M., Scalise, E., Pourtois, G., Afanasev, V.V., Stesmans, A., 2D Materials, 1[2] 2014, 021004. https://doi.org/10.1088/2053-1583/1/2/021004

[179] Cai, B., Zhang, S., Hu, Z., Hu, Y., Zou, Y., Zeng, H., Physical Chemistry Chemical Physics, 17[19] 2015, 12634-12638. https://doi.org/10.1039/C5CP00563A

[180] Mortazavi B., Rahaman O., Makaremi M., Dianat A., Cuniberti G., Rabczuk T., Physica E, 87, 2017, 228-232. https://doi.org/10.1016/j.physe.2016.10.047

[181] Mojumder S., Amin A.A., Islam M.M., Journal of Applied Physics, 118[12] 2015, 124305. https://doi.org/10.1063/1.4931572

[182] Barraza-Lopez S., Synthetic Metals, 210, 2015, 32-41. https://doi.org/10.1016/j.synthmet.2015.06.025

[183] Modarresi M., Kakoee A., Mogulkoc Y., Roknabadi M.R., Computational Materials Science, 101, 2015, 164-167. https://doi.org/10.1016/j.commatsci.2015.01.039

[184] Ribeiro-Soares J., Almeida R.M., Cançado L.G., Dresselhaus M.S., Jorio A., Physical Review B, 91[20] 2015, 205421. https://doi.org/10.1103/PhysRevB.91.205421

[185] Mortazavi B., Rahaman O., Makaremi M., Dianat A., Cuniberti G., Rabczuk T., Physica E, 87, 2017, 228-232. https://doi.org/10.1016/j.physe.2016.10.047

[186] John R., Merlin B., Crystal Structure Theory and Applications, 5, 2016, 43–55. https://doi.org/10.4236/csta.2016.5304

[187] Tao L., Yang C., Wu L., Han L., Song Y., Wang S., Lu P., Modern Physics Letters B, 30[12] 2016, 1650146. https://doi.org/10.1142/S0217984916501463

[188] Kadioglu Y., Ersan F., Gökoğlu G., Aktürk O.Ü., Aktürk E., Materials Chemistry and Physics, 180, 2016, 326-331. https://doi.org/10.1016/j.matchemphys.2016.06.014

[189] Mortazavi B., Dianat A., Cuniberti G., Rabczuk T., Electrochimica Acta, 213, 2016, 865-870. https://doi.org/10.1016/j.electacta.2016.08.027

[190] Dastmard M., Ansari R., Rouhi S., Journal of Molecular Modeling, 27[6] 2021, 163. https://doi.org/10.1007/s00894-021-04776-3

[191] Das S., Mojumder S., Rakib T., Islam M.M., Motalab M., Physica B, 553, 2019, 127-136. https://doi.org/10.1016/j.physb.2018.10.035

[192] Rahman M.H., Chowdhury E.H., Redwan D.A., Mitra S., Hong S., Physical Chemistry Chemical Physics, 23[9] 2021, 5244-5253. https://doi.org/10.1039/D0CP06426B

[193] Shodja H.M., Ojaghnezhad F., Etehadieh A., Tabatabaei M., Mechanics of Materials, 110, 2017, 1-15. https://doi.org/10.1016/j.mechmat.2017.04.001

[194] Mahata A., Mukhopadhyay T., Physical Chemistry Chemical Physics, 20[35] 2018, 22768-22782. https://doi.org/10.1039/C8CP03892A

[195] Khan A.I., Paul R., Subrina S., RSC Advances, 7[80] 2017, 50485-50495. https://doi.org/10.1039/C7RA09209A

[196] Shen L., Lan M., Zhang X., Xiang G., RSC Advances, 7[16] 2017, 9840-9846. https://doi.org/10.1039/C6RA28155A

[197] Zhu Z., Sun Q., Jia Y., AIP Advances, 6[3] 2016, 035012. https://doi.org/10.1063/1.4944621

[198] Wang G., Pandey R., Karna S.P., ACS Applied Materials and Interfaces, 7[21] 2015, 11490-11496. https://doi.org/10.1021/acsami.5b02441

[199] Ji, J., Song, X., Liu, J., Yan, Z., Huo, C., Zhang, S., Su, M., Liao, L., Wang, W., Ni, Z., Hao, Y., Zeng, H., Nature Communications, 7, 2016, 13352. https://doi.org/10.1038/ncomms13352

[200] Gibaja, C., Rodriguez-San-Miguel, D., Ares, P., Gómez-Herrero, J., Varela, M., Gillen, R., Maultzsch, J., Hauke, F., Hirsch, A., Abellán, G., Zamora, F., Angewandte Chemie, 55[46] 2016, 14345-14349. https://doi.org/10.1002/anie.201605298

[201] Aktürk O.U., Özçelik V.O., Ciraci S., Physical Review B, 91[23] 2015, 235446. https://doi.org/10.1103/PhysRevB.91.235446

[202] Zhao M., Zhang X., Li L., Scientific Reports, 5, 2015, 16108.
https://doi.org/10.1038/srep16108

[203] Yang L.F., Song Y., Mi W.B., Wang X.C., Applied Physics Letters, 109[2] 2016,
022103. https://doi.org/10.1063/1.4958702

[204] Le M.Q., Nguyen H.T., Bui T.L., Mechanics of Advanced Materials and Structures,
29[26] 2022, 4993-5005. https://doi.org/10.1080/15376494.2021.1945172

[205] Le M.Q., Computational Materials Science, 201, 2022, 110854.
https://doi.org/10.1016/j.commatsci.2021.110854

[206] Liu G., Gao Z., Zhou J., Physica E, 112, 2019, 59-65.
https://doi.org/10.1016/j.physe.2019.04.002

[207] Aghdasi P., Yousefi S., Ansari R., Journal of Molecular Modeling, 27[1] 2021, 15.
https://doi.org/10.1007/s00894-020-04604-0

[208] Jafari M., Ansari R., Rouhi S., Applied Physics A, 126[2] 2020, 125.
https://doi.org/10.1007/s00339-019-3273-6

[209] Aghdasi P., Yousefi S., Ansari R., Materials Science and Engineering B, 271, 2021,
115219. https://doi.org/10.1016/j.mseb.2021.115219

[210] Mozvashi S.M., Vishkayi S.I., Tagani M.B., Physica E, 118, 2020, 113914.
https://doi.org/10.1016/j.physe.2019.113914

[211] Kripalani D.R., Kistanov A.A., Cai Y., Xue M., Zhou K., Physical Review B, 98[8]
2018, 085410. https://doi.org/10.1103/PhysRevB.98.085410.
https://doi.org/10.1016/j.ssc.2020.113849

[212] Aghdasi P., Ansari R., Solid State Communications, 311, 2020, 113849.

[213] Cetin M., Kirca M., Physical Chemistry Chemical Physics, 25[34] 2023, 22626-
22643. https://doi.org/10.1039/D3CP02026F

[214] Akash T.S., Subad R.A.S.I., Bose P., Islam M.M., Physical Chemistry Chemical
Physics, 23[10] 2021, 6241-6251. https://doi.org/10.1039/D0CP05563H

[215] Singh, D., Gupta, S.K., Lukačević, I., Sonvane, Y., RSC Advances, 6[10] 2016,
8006-8014. https://doi.org/10.1039/C5RA25773E

[216] Le M.Q., Nguyen H.T., Bui T.L., Mechanics of Advanced Materials and Structures,
29[26] 2022, 4993-5005. https://doi.org/10.1080/15376494.2021.1945172

[217] Le M.Q., Computational Materials Science, 201, 2022, 110854.
https://doi.org/10.1016/j.commatsci.2021.110854

[218] Kamal, C., Ezawa, M., Physical Review B, 91[8] 2015, 085423.
https://doi.org/10.1103/PhysRevB.91.085423

[219] Wang, C., Xia, Q., Nie, Y., Rahman, M., Guo, G., AIP Advances, 6[3] 2016,
035204. https://doi.org/10.1063/1.4943548

[220] Chaves, A., Mayers, M.Z., Peeters, F.M., Reichman, D.R., Physical Review B, 93[11] 2016, 115314. https://doi.org/10.1103/PhysRevB.93.115314

[221] Cao, H., Yu, Z., Lu, P., Superlattices and Microstructures, 86, 2015, 501-507. https://doi.org/10.1016/j.spmi.2015.08.006

[222] Guo, D., Shao, B., Li, C., Ma, Y., Superlattices and Microstructures, 100, 2016, 324-334. https://doi.org/10.1016/j.spmi.2016.09.044

[223] Le M.Q., Nguyen H.T., Bui T.L., Mechanics of Advanced Materials and Structures, 29[26] 2022, 4993-5005. https://doi.org/10.1080/15376494.2021.1945172

[224] Le M.Q., Computational Materials Science, 201, 2022, 110854. https://doi.org/10.1016/j.commatsci.2021.110854

[225] Akbari O., Ansari R., Rouhi S., Materials Research Express, 5[1] 2018, 015025. https://doi.org/10.1088/2053-1591/aaa217

[226] Yang B., Li M., Wang J., Zhang J., Liao D., Yue Y., Physical Chemistry Chemical Physics, 21[23] 2019, 12372-12379. https://doi.org/10.1039/C9CP01605H

[227] Guo J., Liu G., Sun Z., Journal of Wuhan University of Technology - Materials Science, 39[1] 2024, 206-212. https://doi.org/10.1007/s11595-024-2873-y

[228] Aghdasi P., Ansari R., Rouhi S., Goli M., Gilakjani H.A., Physica B, 574, 2019, 411672. https://doi.org/10.1016/j.physb.2019.411672

[229] Chen B., Xue L., Han Y., Li X.Q., Yang Z., Materials Today Communications, 22, 2020, 100791. https://doi.org/10.1016/j.mtcomm.2019.100791

[230] Shang Y., Wang M., Hu Q., Gao Q., Xin Y., Shi H., Xiao Z., Chu P.K., Huang A., Computational Condensed Matter, 23, 2020, e00473. https://doi.org/10.1016/j.cocom.2020.e00473

[231] Aghdasi P., Ansari R., Yousefi S., Goli M., Superlattices and Microstructures, 139, 2020, 106414. https://doi.org/10.1016/j.spmi.2020.106414

[232] Yousefi S., Ansari R., Aghdasi P., Mozvashi S.M., Physica E, 124, 2020, 114349. https://doi.org/10.1016/j.physe.2020.114349

[233] Aghdasi P., Ansari R., Rouhi S., Yousefi S., Journal of Molecular Graphics and Modelling, 101, 2020, 107725. https://doi.org/10.1016/j.jmgm.2020.107725

[234] Liu G., Gao Z., Zhou J., Physica E, 112, 2019, 59-65. https://doi.org/10.1016/j.physe.2019.04.002

[235] Kou, L., Chen, C., Smith, S.C., Journal of Physical Chemistry Letters, 6[14] 2015, 2794-2805. https://doi.org/10.1021/acs.jpclett.5b01094

[236] Ribeiro-Soares, J., Almeida, R.M., Cançado, L.G., Dresselhaus, M.S., Jorio, A., Physical Review B, 91[20] 2015, 205421. https://doi.org/10.1103/PhysRevB.91.205421

[237] Köpf, M., Eckstein, N., Pfister, D., Grotz, C., Krüger, I., Greiwe, M., Hansen, T., Kohlmann, H., Nilges, T., Journal of Crystal Growth, 405, 2014, 6-10. https://doi.org/10.1016/j.jcrysgro.2014.07.029

[238] Serrano-Ruiz, M., Caporali, M., Ienco, A., Piazza, V., Heun, S., Peruzzini, M., Advanced Materials Interfaces, 3[3] 2016, 1500441. https://doi.org/10.1002/admi.201500441

[239] Mu, Y., Si, M.S., EPL, 112[3] 2015, 37003. https://doi.org/10.1209/0295-5075/112/37003

[240] Wang, L., Sofer, Z., Pumera, M., ChemElectroChem, 2[3] 2015, 324-327. https://doi.org/10.1002/celc.201402363

[241] Zeng, J., Cui, P., Zhang, Z., Physical Review Letters, 118[4] 2017, 046101. https://doi.org/10.1103/PhysRevLett.118.046101

[242] Guan, J., Zhu, Z., Tománek, D., ACS Nano, 8[12] 2014, 12763-12768. https://doi.org/10.1021/nn5059248

[243] Wu, M., Fu, H., Zhou, L., Yao, K., Zeng, X.C., Nano Letters, 15[5] 2015, 3557-3562. https://doi.org/10.1021/acs.nanolett.5b01041

[244] Zhao, T., He, C.Y., Ma, S.Y., Zhang, K.W., Peng, X.Y., Xie, G.F., Zhong, J.X., Journal of Physics - Condensed Matter, 27[26] 2015, 265301. https://doi.org/10.1088/0953-8984/27/26/265301

[245] Sha, Z.D., Pei, Q.X., Ding, Z., Jiang, J.W., Zhang, Y.W., Journal of Physics D, 48[39] 2015, 395303. https://doi.org/10.1088/0022-3727/48/39/395303

[246] Sha Z.D., Pei Q.X., Ding Z., Jiang J.W., Zhang Y.W., Journal of Physics D, 48[39] 2015, 395303. https://doi.org/10.1088/0022-3727/48/39/395303

[247] Hao, F., Chen, X., Journal of Applied Physics, 118[23] 2015, 234304. https://doi.org/10.1063/1.4937920

[248] Liu, N., Hong, J., Pidaparti, R., Wang, X., Nanoscale, 8[10] 2016, 5728-5736. https://doi.org/10.1039/C5NR08682E

[249] Sorkin, V., Zhang, Y.W., Nanotechnology, 27[39] 2016, 395701. https://doi.org/10.1088/0957-4484/27/39/395701

[250] Sorkin, V., Zhang, Y.W., 2D Materials, 2[3] 2015, 035007. https://doi.org/10.1088/2053-1583/2/3/035007

[251] Xu, F., Ma, H., Lei, S., Sun, J., Chen, J., Ge, B., Zhu, Y., Sun, L., Nanoscale, 8[28] 2016, 13603-13610. https://doi.org/10.1039/C6NR02487D

[252] Kou, L., Ma, Y., Smith, S.C., Chen, C., Journal of Physical Chemistry Letters, 6[9] 2015, 1509-1513. https://doi.org/10.1021/acs.jpclett.5b00522

[253] Wang, G., Loh, G.C., Pandey, R., Karna, S.P., Nanotechnology, 27[5] 2015, 055701. https://doi.org/10.1088/0957-4484/27/5/055701

[254] Hu, T., Han, Y., Dong, J., Nanotechnology, 25[45] 2014, 455703. https://doi.org/10.1088/0957-4484/25/45/455703

[255] Wang, L., Kutana, A., Zou, X., Yakobson, B.I., Nanoscale, 7[21] 2015, 9746-9751. https://doi.org/10.1039/C5NR00355E

[256] Sorkin, V., Zhang, Y.W., Nanotechnology, 26[23] 2015, 235707. https://doi.org/10.1088/0957-4484/26/23/235707

[257] Verma, D., Hourahine, B., Frauenheim, T., James, R.D., Dumitric, Ă.T., Physical Review B, 94[12] 2016, 121404. https://doi.org/10.1103/PhysRevB.94.121404

[258] Wang, H., Li, X., Li, P., Yang, J., Nanoscale, 9[2] 2017, 850-855. https://doi.org/10.1039/C6NR08550D

[259] Le M.Q., Nguyen H.T., Bui T.L., Mechanics of Advanced Materials and Structures, 29[26] 2022, 4993-5005. https://doi.org/10.1080/15376494.2021.1945172

[260] Le M.Q., Computational Materials Science, 201, 2022, 110854. https://doi.org/10.1016/j.commatsci.2021.110854

[261] Liu G., Gao Z., Zhou J., Physica E, 112, 2019, 59-65. https://doi.org/10.1016/j.physe.2019.04.002

[262] Liu P., Pei Q.X., Huang W., Zhang Y.W., Journal of Physics D, 50[48] 2017, 485303. https://doi.org/10.1088/1361-6463/aa8f66

[263] Liu N., Hong J., Zeng X., Pidaparti R., Wang X., Physical Chemistry Chemical Physics, 19[20] 2017, 13083-13092. https://doi.org/10.1039/C7CP01033H

[264] Liu N., Hong J., Pidaparti R., Wang X., Nanoscale, 8[10] 2016, 5728-5736. https://doi.org/10.1039/C5NR08682E

[265] Garrison E., Chan C.K., Peng X., AIP Advances, 8[11] 2018, 115124. https://doi.org/10.1063/1.5054619

[266] Zhang B., Zhang L., Yang N., Zhao X., Chen C., Cheng Y., Rasheed I., Ma L., Zhang J., Journal of Physical Chemistry C, 126[2] 2022, 1094-1098. https://doi.org/10.1021/acs.jpcc.1c10187

[267] Le M.Q., Nanotechnology, 29[19] 2018, 195701. https://doi.org/10.1088/1361-6528/aaaacf

[268] Hao F., Chen X., Journal of Applied Physics, 118[23] 2015, 234304. https://doi.org/10.1063/1.4937920

[269] Sorkin V., Zhang Y.W., Nanotechnology, 27[39] 2016, 395701. https://doi.org/10.1088/0957-4484/27/39/395701

[270] Nguyen D.T., Le M.Q., Nguyen V.T., Bui T.L., Superlattices and Microstructures, 112, 2017, 186-199. https://doi.org/10.1016/j.spmi.2017.09.021

[271] Hao F., Liao X., Li M., Xiao H., Chen X., Journal of Physics Condensed Matter, 30[31] 2018, 315302. https://doi.org/10.1088/1361-648X/aacfd1

[272] Zhang B., Zhang L., Wang Z., Li Y., Cheng Y., Ma L., Zhang J., Journal of Materials Chemistry A, 9[24] 2021, 13855-13860. https://doi.org/10.1039/D1TA02595C

[273] Sahoo G., Biswal A., Physica B, 697, 2025, 416742. https://doi.org/10.1016/j.physb.2024.416742

[274] Wang H., Li X., Li P., Yang J., Nanoscale, 9[2] 2017, 850-855. https://doi.org/10.1039/C6NR08550D

[275] Ansari R., Aghdasi P., Shahnazari A., Applied Physics A, 130[4] 2024, 214. https://doi.org/10.1007/s00339-024-07371-4

[276] Santra P., Ghaderzadeh S., Ghorbani-Asl M., Komsa H.P., Besley E., Krasheninnikov A.V., npj 2D Materials and Applications, 8[1] 2024, 33. https://doi.org/10.1038/s41699-024-00472-x

[277] Aghdasi P., Shahnazari A., Ansari R., Applied Physics A, 130[4] 2024, 219. https://doi.org/10.1007/s00339-024-07371-4

[278] Chang X., Yin N., Li H., Materials Research Express, 6[11] 2019, 115043. https://doi.org/10.1088/2053-1591/ab4661

[279] Sorkin V., Zhang Y.W., Extreme Mechanics Letters, 14, 2017, 2-9. https://doi.org/10.1016/j.eml.2016.10.011

[280] Liu N., Pidaparti R., Wang X., Physical Chemistry Chemical Physics, 20[13] 2018, 8668-8675. https://doi.org/10.1039/C7CP08540K

[281] Li T., Bi X.Y., Kong J.W., Acta Physica Sinica, 72[12] 2023, 126201. https://doi.org/10.7498/aps.72.20230084

[282] Cai, Y., Ke, Q., Zhang, G., Yakobson, B.I., Zhang, Y.W., Journal of the American Chemical Society, 138[32] 2016, 10199-10206. https://doi.org/10.1021/jacs.6b04926

[283] Sha, Z.D., Pei, Q.X., Zhang, Y.Y., Zhang, Y.W., Nanotechnology, 27[31] 2016, 315704. https://doi.org/10.1088/0957-4484/27/31/315704

[284] Hao F., Chen X., Journal of Applied Physics, 120[16] 2016, 165104. https://doi.org/10.1063/1.4966167

[285] Li X., Ma L., Wang D., Zeng X.C., Wu X., Yang J., Nanoscale, 8[41] 2016, 17801-17808. https://doi.org/10.1039/C6NR05414E

[286] Sorkin V., Zhang Y.W., Nanotechnology, 28[7] 2017, 075704. https://doi.org/10.1088/1361-6528/aa537b

[287] Aktürk, E., Aktürk, O.Ü., Ciraci, S., Physical Review B, 94[1], 2016, 014155. https://doi.org/10.1103/PhysRevB.94.014115

[288] Ji, W.X., Zhang, C.W., Ding, M., Zhang, B.M., Li, P., Li, F., Ren, M.J., Wang, P.J., Zhang, R.W., Hu, S.J., Yan, S.S., New Journal of Physics, 18[8], 2016, 083002. https://doi.org/10.1088/1367-2630/18/8/083002

[289 Walker, E.S., Na, S.R., Jung, D., March, S.D., Kim, J.S., Trivedi, T., Li, W., Tao, L., Lee, M.L., Liechti, K.M., Akinwande, D., Bank, S.R., Nano Letters, 16[11], 2016, 6931-6938. https://doi.org/10.1021/acs.nanolett.6b02931

[290] Le M.Q., Nguyen H.T., Bui T.L., Mechanics of Advanced Materials and Structures, 29[26] 2022, 4993-5005. https://doi.org/10.1080/15376494.2021.1945172

[291] Le M.Q., Computational Materials Science, 201, 2022, 110854. https://doi.org/10.1016/j.commatsci.2021.110854

[292] Aghdasi P., Ansari R., Rouhi S., Goli M., Superlattices and Microstructures, 135, 2019, 106242. https://doi.org/10.1016/j.spmi.2019.106242

[293] Li Z., Pan H., Chu H., Pan Z., Li Y., Zhao S., Li D., Physical Review B, 110[12] 2024, 125429. https://doi.org/10.1103/PhysRevB.110.125429

[294] Chowdhury E.H., Rahman M.H., Bose P., Jayan R., Islam M.M., Physical Chemistry Chemical Physics, 22[48] 2020. 28238-28255. https://doi.org/10.1039/D0CP04785F

[295] Jafari M., Ansari R., Rouhi S., Superlattices and Microstructures, 140, 2020, 106476. https://doi.org/10.1016/j.spmi.2020.106476

[296] Mannix, A.J., Zhou, X.F., Kiraly, B., Wood, J.D., Alducin, D., Myers, B.D., Liu, X., Fisher, B.L., Santiago, U., Guest, J.R., Yacaman, M.J., Ponce, A., Oganov, A.R., Hersam, M.C., Guisinger, N.P., Science, 350[6267] 2015, 1513-1516. https://doi.org/10.1126/science.aad1080

[297] Piazza, Z.A., Hu, H.S., Li, W.L., Zhao, Y.F., Li, J., Wang, L.S., Nature Communications, 5, 2014, 3113. https://doi.org/10.1038/ncomms4113

[298] Li, W.L., Chen, Q., Tian, W.J., Bai, H., Zhao, Y.F., Hu, H.S., Li, J., Zhai, H.J., Li, S.D., Wang, L.S., Journal of the American Chemical Society, 136[35] 2014, 12257-12260. https://doi.org/10.1021/ja507235s

[299] Li, W.L., Zhao, Y.F., Hu, H.S., Li, J., Wang, L.S., Angewandte Chemie – International, 53[22] 2014, 5540-5545. https://doi.org/10.1002/anie.201402488

[300] Zabolotskiy, A.D., Lozovik, Y.E., Physical Review B, 94[16] 2016, 165403. https://doi.org/10.1103/PhysRevB.94.165403

[301] Lopez-Bezanilla, A., Littlewood, P.B., Physical Review B, 93[24] 2016, 241405. https://doi.org/10.1103/PhysRevB.93.241405

[302] Xu, S., Zhao, Y., Liao, J., Yang, X., Xu, H., Nano Research, 9[9] 2016, 2616-2622. https://doi.org/10.1007/s12274-016-1148-0

[303] Zhong C., Sun M., Altalhi T., Yakobson B.I., Materials, 17[9] 2024, 1967. https://doi.org/10.3390/ma17091967

[304] Yuan, J., Zhang, L.W., Liew, K.M., RSC Advances, 5[91] 2015, 74399-74407. https://doi.org/10.1039/C5RA14939H

[305] Wang, H., Li, Q., Gao, Y., Miao, F., Zhou, X.F., Wan, X.G., New Journal of Physics, 18[7] 2016, 073016. https://doi.org/10.1088/1367-2630/18/7/073016

[306] Le, M.Q., Mortazavi, B., Rabczuk, T., Nanotechnology, 27[44] 2016, 445709. https://doi.org/10.1088/0957-4484/27/44/445709

[307] Pang Z., Qian X., Wei Y., Yang R., EPL, 116[3] 2016, 36001. https://doi.org/10.1209/0295-5075/116/36001

[308] Zhang Z., Mannix A.J., Hu Z., Kiraly B., Guisinger N.P., Hersam M.C., Yakobson B.I., Nano Letters, 16[10] 2016, 6622-6627. https://doi.org/10.1021/acs.nanolett.6b03349

[309] Yuan J., Yu N., Xue K., Miao X., RSC Advances, 7[14] 2017, 8654-8660. https://doi.org/10.1039/C6RA28454J

[310] Wang X., Wu R., Xu T., Gao Y., Materials Research Express, 8[6] 2021, 065003. https://doi.org/10.1088/2053-1591/ac04ec

[311] Sha Z.D., Pei Q.X., Zhou K., Dong Z., Zhang Y.W., Extreme Mechanics Letters, 19, 2018, 39-45. https://doi.org/10.1016/j.eml.2017.12.008

[312] Saaoud M., Sadki K., Drissi L.B., Djeffal F., EPJ Applied Physics, 90, 2020, 2020011. https://doi.org/10.1051/epjap/2020200119

[313] Pham V.T., Fang T.H., Thin Solid Films, 709, 2020, 138197. https://doi.org/10.1016/j.tsf.2020.138197

[314] Shao L., Li Y., Yuan Q., Li M., Du Y., Zeng F., Ding P., Ye H., Materials Research Express, 4[4] 2017, 045020. https://doi.org/10.1088/2053-1591/aa6bba

[315] Abadi R., Jenabidehkordi A., Rabczuk T., Computational Materials Science, 178, 2020, 109625. https://doi.org/10.1016/j.commatsci.2020.109625

[316] Giannopoulos G.I., Computational Materials Science, 129, 2017, 304-310. https://doi.org/10.1016/j.commatsci.2016.12.045

[317] Mortazavi B., Rahaman O., Dianat A., Rabczuk T., Physical Chemistry Chemical Physics, 18[39] 2016, 27405-27413. https://doi.org/10.1039/C6CP03828J

[318] Han T.W., Li X.Z., Zhao Z.R., Gu Y.T., Ma C., Zhang X.Y., Acta Physica Sinica, 73[11] 2024, 116201. https://doi.org/10.7498/aps.73.20240066

[319] Zhang Z., Yang Y., Penev E.S., Yakobson B.I., Advanced Functional Materials, 27[9] 2017, 1605059. https://doi.org/10.1002/adfm.201605059

[320] Wang V., Geng W.T., Journal of Physical Chemistry C, 121[18] 2017, 10224-10232. https://doi.org/10.1021/acs.jpcc.7b02582

[321] Tang X., Chen H., Ding Y., Materials Research Express, 6[3] 2019, 035010. https://doi.org/10.1088/2053-1591/aaf367

[322] Vajary R.A., Tagani M.B., Vishkayi S.I., Modern Physics Letters B, 32[28] 2018, 1850347. https://doi.org/10.1142/S0217984918503475

[323] Peköz R., Konuk M., Kilic M.E., Durgun E., ACS Omega, 3[2] 2018, 1815-1822. https://doi.org/10.1021/acsomega.7b01730

[324] Dethan J.F.N., Physical Chemistry Chemical Physics, 23[31] 2021, 17009-17017. https://doi.org/10.1039/D1CP01831K

[325] Wang Z., Lü T.Y., Wang H.Q., Feng Y.P., Zheng J.C., Physical Chemistry Chemical Physics, 18[46] 2016, 31424-31430. https://doi.org/10.1039/C6CP06164H

[326] Das D.K., Sarkar J., Singh S.K., Computational Materials Science, 151, 2018, 196-203. https://doi.org/10.1016/j.commatsci.2018.05.006

[327] Das D.K., Kumar B., Computational Materials Science, 233, 2024, 112678. https://doi.org/10.1016/j.commatsci.2023.112678

[328] Das D.K., Kumar B., Materials Today Communications, 40, 2024, 110142. https://doi.org/10.1016/j.mtcomm.2024.110142

[329] Mahdavifar S., Shayesteh S.F., Tagani M.B., Physica E, 134, 2021, 114837. https://doi.org/10.1016/j.physe.2021.114837